Practices
5

for
Orchestrating
Productive
Task-Based
Discussions in
Science

Jennifer L. Cartier
University of Pittsburgh

Margaret S. Smith
University of Pittsburgh

Mary Kay Stein
University of Pittsburgh

Danielle K. Ross
University of Pittsburgh

NATIONAL COUNCIL OF
TEACHERS OF MATHEMATICS

Copyright © 2013 by
The National Council of Teachers of Mathematics, Inc.
1906 Association Drive, Reston, VA 20191-1502
(703) 620-9840; (800) 235-7566; www.nctm.org
All rights reserved

ISBN: 978-0-87353-745-2

The Cataloging-in-Publication data is on file with the Library of Congress.

The National Council of Teachers of Mathematics is the public voice of mathematics education, supporting teachers to ensure equitable mathematics learning of the highest quality for all students through vision, leadership, professional development, and research.

Printed in the United States of America

Contents

CHAPTER 6

CHAPTER 7

Foreword

Accomplished science teachers are a joy to watch in the classroom. They pose engaging questions, they use students' ideas and experiences as stepping-stones for knowledge construction by the whole class, and they allow students to test their emerging theories through experiment and conversation. They do all this while pressing young learners, gently but relentlessly, to develop coherent and evidence-based explanations of the natural world.

But what often appears to be a spontaneous teaching performance is really the product of careful planning. The classroom activity itself is guided by instructional goals and principles for in-the-moment decision making that may not be evident to an observer. In fact, much of what contributes to expertise is invisible. For those of us who are passionate about the continual improvement of teaching, we want to know what is going on "below the surface" of classroom activity that allows some educators to consistently open up opportunities for learning, and to do so for all students. The authors of this book have clearly been hard at work on this puzzle.

We too, at the University of Washington, have been interested in the trajectories of teachers' practice over time, and the transitions from mere survival in the classroom, to competence, to instructional excellence. Several years ago we followed dozens of novice educators from our teacher preparation program into their classrooms and observed how they planned lessons and interacted with students. We found, in one case after another, that our young teachers struggled to manage instructional conversations with students. It wasn't that they could not get talk started, as they were quite able to get it under way. For example, one of our teachers asked his sophomore biology students why they thought physical traits in humans sometimes appeared to skip a generation ("Who has their grandfather's nose?"); a middle school teacher asked her students to speculate why some very heavy objects would float in water (ships) while some not-so-heavy things (a grain of sand) would sink. These questions initiated a lot of hypothesizing, but in a matter of minutes our novices were responding to students with remarks like "OK," "Uh-huh," "That's interesting," and "Shall we make a list of these ideas?" When we interviewed them after class they reflected on their discussions in the same way one might talk about a hiking trip gone awry: "I started us off OK but then lost track of where we were going," "The kids went in directions I was not expecting," or "I had no idea where we were supposed to end up."

This marked the beginning of our research group's focus on classroom talk as a tool for supporting student reasoning. We continued to study our beginning teachers, but we also studied science educators who were highly skilled at talk, and we began to notice that the more expert the educator, the more prominent the students' roles in the conversations were. Experts actually talked less, but they were strategic in how they responded to students' ideas. We also could see that experts at classroom discourse had a clear goal for the talk, even though it seemed they were letting students control much of the dialogue.

It should be clear by now that we and the authors of this book share a passion about classroom discourse, and it is no surprise that there have been some interesting convergences in our ideas. Here's my sense of what we agree on, based on the rich examples of teaching practice in the pages that follow.

First, carefully orchestrated talk promotes deep and robust reasoning. Put more simply, talk mediates thinking, and students need more chances to talk, but with specific forms of guidance. Managing talk is also critical for engaging learners in the characteristic activities of science—that is, specialized forms of language are needed to formulate questions that interest students; to build and critique theories; to collect, analyze, and interpret data; to evaluate hypotheses through experimentation; and to communicate findings. This is actually unnatural talk for students; they need to have modeled for them how one expresses hypotheses in response to observations, how a person argues about evidence, and how they might critique another person's scientific model.

Second, a teacher has to anticipate what kinds of activity and talk will be needed to accomplish particular instructional tasks. But being smart about anticipating means that you interrogate your own understanding of the subject matter. We urge the teachers we work with to base their units of instruction on a complex phenomenon and then, working in groups together, develop a full causal explanation for that event or process before they start planning lessons. The examples in this book portray students who are involved in high cognitive demand activities, and it is clear in the exemplar vignettes that the teacher has deepened her or his understanding of the content to help them interact with students' ideas. These teachers are prepared.

Third, we seem to agree that teaching is "working on students' ideas." This means that teachers have to elicit what students are thinking and make that thinking public and visible in some form. They need to ask students to compare ideas, critique the ideas of others, change ideas in response to new experiences and concepts, be able to identify where the gaps in their current understandings are, and identify resources that will move their thinking forward.

And finally, great teaching is "learnable," especially if it can be represented as principled practices and if teachers get chances to try out these strategies with students over an extended period of time. This is the utility of having a focused set of teaching practices, such as the five in this book, that can be enacted with any kind of science subject matter—you get better at them with repeated and varied attempts. In our experience, students also get better at the discourses, and in the process they adjust to the higher expectations for intellectual work.

Occasionally when I help teachers work on their practice, I reflect back on my own thirteen years of middle school teaching. I was a popular educator—I knew my subject matter, I was organized, and I enjoyed working with young learners. But at that time there was little known about the power of talk in classrooms or about being mindful of something called cognitive demand. At that time you just taught what was in the curriculum! When teachers in my school attended professional development, it was generic and rarely (as it remains today) about effective teaching practice. Instructional excellence remained an idea, an aspiration, and it was never embodied in rich classroom examples of rigorous and responsive interactions with students.

I think that back then, if I had access to the ideas in *5 Practices for Orchestrating Productive Task-Based Discussions in Science,* I might have marched down to the principal's office and asked that our science department create its own professional development experiences, using the ideas in this book as a framework. If you think similarly after reading this book, here are some steps that we've seen teacher groups follow that helps them to structure their work together. They begin with a self-study, spending a few weeks observing one another's practice and doing a quick analysis of the patterns of talk that currently characterize their classrooms. They sometimes look at the kinds of work that students produce as a result of classroom talk to use as a baseline for later comparisons. They set some goals for experimenting with new ways of teaching. They try to take "first steps," changing some aspects of their classroom discourse, giving it a try three or four times before assessing where they were making headway and what adjustments are needed. These teachers make good use of video to capture conversations with students, and they play it back later to ask each other, "What were my students thinking here, and what might I have done differently to challenge them?"

Today the basic ingredients for advancing one's practice are often ready at hand. You need willing colleagues, a cooperative administrator to give you time to meet, access to some technology, and perhaps most importantly, you need a set of ideas to work on together. The illuminating vignettes, generative frameworks, and helpful tools in this book are a great set of resources to support you in this journey.

Mark A. Windschitl
Professor of Curriculum and Instruction,
University of Washington

Preface

In this book, we present and discuss a framework for orchestrating productive discussions in science that are rooted in student thinking and that emerge from students' work on demanding tasks. Such tasks provide opportunities for students to engage in the disciplinary practices described in the Next Generation Science Standards (Achieve, Inc. 2013) while also developing understanding of key patterns and/or concepts in science. The framework presented throughout the book identifies a set of instructional practices that will help teachers effectively use student work as the launching point for discussions in which students address important science ideas, consider alternative explanations, identify contradictions between evidence and claims, and develop or consolidate understandings of new concepts. The premise underlying the book is that the identification and use of a codified set of practices can make student-centered approaches to science instruction accessible to and manageable for more teachers. By giving teachers a road map of things that they can do in advance and during whole-class discussions, these practices have the potential for helping them to more effectively orchestrate discussions that are responsive to both students' thinking and core practices and ideas within science disciplines.

Throughout the book, we illustrate the instructional practices with episodes that take the reader inside science classrooms. In particular, we make significant use of three narrative cases: the Cases of Kelly Davis, Nathan Gates, and Kendra Nichols. We introduce the Cases of Kelly Davis and Nathan Gates in chapter 2 to contrast the quality of instruction that does and does not utilize the Five Practices framework. We explore the Case of Kendra Nichols in considerable depth in chapters 3 and 4 as each of the five practices is examined in detail, and refer to it again in subsequent chapters as we consider broader issues related to integrating the five practices into everyday instruction. These cases, and other vignettes that appear in the book, are based on real events and are intended to make salient certain types of teacher-student interactions and the level and type of thinking required to teach with understanding. As such, these episodes of teaching reflect what we have observed, and they should be thought of as composites that have been enhanced at times in order to bring out specific aspects of instruction we wish to highlight.

Following research that has established the importance of learners' construction of their own knowledge (Bransford, Brown, and Cocking 2000), we have designed this book to encourage the active engagement of readers. In several places, we have provided notes (titled "Active Engagement") that suggest ways in which the reader can engage with specific artifacts of classroom practice (e.g., narrative cases of classroom instruction, transcripts of classroom interactions, instructional tasks, or samples of student work). Rather than passively read the book from cover to cover, readers are encouraged to take our suggestions to heart and pause for a moment to grapple with the information in the ways suggested. By actively processing the information, readers' understandings will be deepened, as will their ability to access and use the knowledge flexibly in their own professional work. In addition, within some chapters we have provided suggestions (titled "Try This!") regarding how teachers can explore the ideas from a chapter in their own classrooms.

Although the primary focus of the book (chapters 2, 3, 4, and 7) is the Five Practices model first established in *5 Practices for Orchestrating Productive Mathematics Discussions* (Smith and Stein 2011), we also explore other issues that support teachers' ability to orchestrate productive classroom discussions. Specifically, in chapter 1 we emphasize the need to set clear goals for what students will learn as a result of instruction and to identify a task that is consistent with those learning goals prior to engaging in the five practices. In chapter 5 we focus explicitly on the types of questions that teachers can ask to challenge students' thinking and the moves that teachers can make to promote the participation of students in whole-class discussions. We situate the Five Practices model for facilitating a discussion within the broader context of instructional design in chapter 6. The book concludes with chapter 7, in which we describe the lessons learned by beginning secondary science teachers as they endeavored to conduct task-based discussions in science using the five practices.

Acknowledgments

For their help in the creation of this book, the authors would like to thank:

- The Knowles Science Teaching Foundation (KSTF), which has provided opportunities to share and refine the Five Practices model with motivated, talented, and reflective teachers.
- Michele Cheyne, KSTF Biology Teacher Developer and former science education faculty member at the University of Pittsburgh, who has collaborated on efforts to disseminate the Five Practices model to novice teachers and provided crucial feedback during the development of this book.
- Brittany Barickman, Natalie Dutrow, Kristin Germinario, Nicole Lien, Sarah Macway, Laura Nutter, Rachel Packer, Emma Ross, Lyudmila Shemyakina, and Helen Snodgrass—KSTF Biology Fellows who have thrown themselves energetically into the implementation of the Five Practices model in their classrooms. Their insights about the model's utility and limitations have been instrumental in developing this book.
- The teacher candidates and teachers with whom we have worked at the University of Pittsburgh and in school districts across the United States. We have been privileged to learn from them as they implemented the Five Practices model and shared their feedback with us.
- Steve Scoville, who provided assistance with illustrations.
- Elaine Lucas Evans and Helen Snodgrass, who contributed instructional tasks.
- Don Smith, who served as a science content consultant and happy hour host following the team's writing retreats.
- Deanna Weber Prine, who provided assistance with the final preparation of the manuscript.

Introduction

In 2013, Achieve, Inc. published the Next Generation Science Standards (NGSS). In this document, science educators draw on decades of student learning research to establish conceptual learning goals and identify key disciplinary practices (see fig. 0.1) that should form the basis of a coherent science education in kindergarten through grade 12. NGSS presents an ambitious vision for science instruction "in which students, over multiple years of school, actively engage in scientific and engineering practices and apply crosscutting concepts to deepen their understanding of the core ideas in these fields" (NRC 2012, pp. 8–9).

Science Practices for K–12 Classrooms

1. Asking questions
2. Developing and using models
3. Planning and carrying out investigations
4. Analyzing and interpreting data
5. Using mathematics and computational thinking
6. Constructing explanations
7. Engaging in argument from evidence
8. Obtaining, evaluating, and communicating information

Fig. 0.1. Science practices for K–12 classrooms. From *The Next Generation Science Standards* (Achieve, Inc. 2013).

Instructional Challenges

Clearly, teachers face many challenges in attempting to make this vision a reality in kindergarten–grade 12 classrooms. One such challenge is the *selection and/or design of instructional tasks* that will provide opportunities for students to learn canonical science ideas while also participating in disciplinary practices. Many readily available science learning tasks enable students to do one or the other—learn science content or engage in disciplinary practices—but not both. Moreover, it is common for tasks to constrain or direct students' work to such a degree that their participation in science practices is merely perfunctory. For example, in task A (fig. 0.2) students graph the data provided in order to identify a pattern (i.e., notice which cities have average temperatures that fall in the desirable range during various times of the year). However, the task does not prompt students to notice differences in the cities' climate patterns or to propose an explanation for this pattern. While task A does ask students to use data to answer a question, it also provides detailed instructions about how to represent and analyze the data. Figure 0.3 shows the "correct" data representation for this task. It is therefore unlikely that students will vary greatly in the way they approach the task and, consequently, there will be little opportunity for them to engage in argument from evidence or motivation for them to communicate information with one another.

Task A

Jeremy is planning ahead for his vacation next year. He has decided that he'd like to travel to a place where he can enjoy outdoor camping, hiking, and fishing with his Labrador retriever, Sadie. Jeremy's tent is rated for temperatures above freezing (32°F). Sadie prefers not to be too active when the temperature is over 70°F.

Create a bar graph that shows the average monthly high and low temperatures in each city. Identify where and when Jeremy should go on vacation.

	Amber Lake		Bakersville		Chesterton	
	Mean Low Temperature °F	Mean High Temperature °F	Mean Low Temperature °F	Mean High Temperature °F	Mean Low Temperature °F	Mean High Temperature °F
January	20	38	40	61	53	80
February	22	42	43	65	54	78
March	28	51	49	72	52	72
April	38	64	56	78	44	68
May	47	73	65	85	35	57
June	56	81	70	90	34	53
July	61	85	73	92	32	50
August	60	83	72	92	34	54
September	52	76	67	88	38	60
October	41	65	57	81	42	65
November	33	53	49	72	52	69
December	24	41	42	63	54	79

Fig. 0.2. Task A, an earth science task for students in grade 6

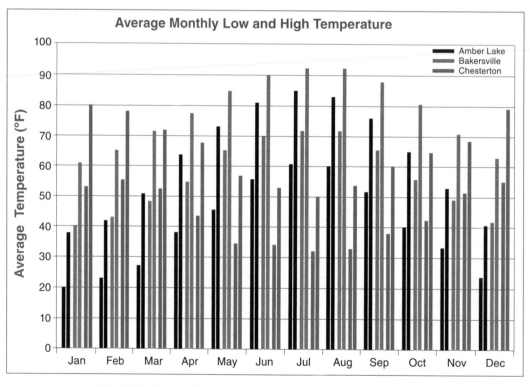

Fig. 0.3. The "correct" way to represent the data as prompted in task A

A teacher who is using task A in the classroom, and who also wishes to enact the vision put forth in the NGSS, would need to modify the task so that students would have a reason to consider the underlying science ideas and an opportunity to reason about those ideas with one another. Task B (fig. 0.4) is an example of such a modification.

Task B

Jeremy is planning ahead for his vacation next year. He has decided that he'd like to travel to a place where he can enjoy outdoor camping, hiking, and fishing with his Labrador retriever, Sadie. Jeremy's tent is rated for temperatures above freezing (32°F). Sadie prefers not to be too active when the temperature is over 70°F.

Using the data provided, create a representation that will help you to show which city Jeremy should visit and at what time of year (spring, fall, winter, or summer). You may represent your data in any way you choose. You may choose to represent all or only some of the data, as long as you can use your representation to justify your recommendations for Jeremy's vacation (where to go and when to go there).

Fig. 0.4. Task B prompts students to select and represent data for the purpose of making an argument.

Using the same data set as task A, this modified task prompts students to create a representation *for the purpose of convincing others* of the validity of their recommendation. In order to complete the task, students need to decide what data to use, whether or how to transform the data, and how to represent it. There are many ways in which students could approach the task and provide a reasonable answer that they could justify using the given data. For example, students might compute an average high and low temperature for each season (fig. 0.5); use a bar graph to plot only the high and low temperature values for the cities during months when the temperature range is acceptable (fig. 0.6); or plot the temperature range for each city during every month of the year, including horizontal lines on the graph that indicate the acceptable temperature range for Jeremy's vacation (fig. 0.7).

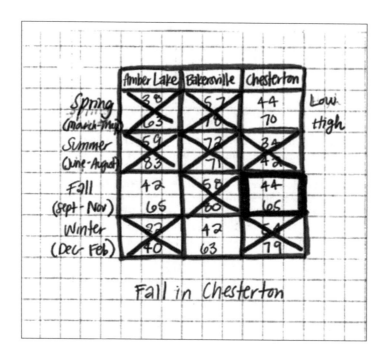

Fig. 0.5. This group of students computed the average high and low temperature in each season for all three cities. They then selected fall in Chesterton as the most desirable time and location for Jeremy.

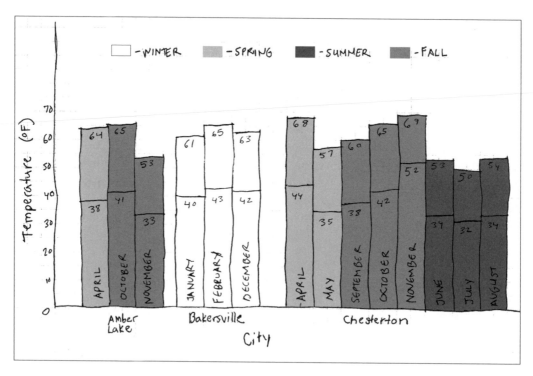

Fig. 0.6. These students identified all the months in each city when the high and low temperatures were within the desirable range. They used a bar graph to plot the data for each of the potential vacation sites and times.

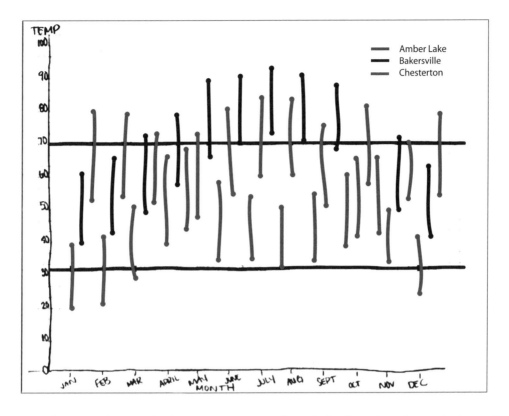

Fig. 0.7. This group chose to plot the temperature range for each city throughout the year. In their oral presentation, they explained that Jeremy could go to any city with a temperature range that falls between the two black horizontal lines (drawn at 32°F and 70°F) during a particular month.

Through a purposefully orchestrated discussion of various representations, the teacher could provide students with an opportunity to notice a key pattern: The Northern and Southern Hemispheres experience the seasons at opposite times of the year. Thus, task B may enable students to engage in science practices while also learning important disciplinary content. In task A, students are not asked to notice or describe this pattern. Instead, they must simply graph the data and name a single location and month where the temperature range falls within Jeremy's requirements.

Task B not only provides students with opportunities to make authentic choices about how to analyze and represent data, but it also requires them to rationalize or defend their answers to the problem of where Jeremy ought to take his vacation. Tasks of this kind can pose a particularly thorny challenge for teachers, that of *enabling and supporting productive discussion that is grounded in students' own work*. More than a decade ago, Cazden noted that the nature of talk in the classroom "can have considerable cognitive or social significance" (2001, p. 53) for students, and she urged teachers and teacher educators to think seriously about promoting equitable and active student engagement in classroom discourse. She also warned that "it is easy to imagine talk in which ideas are explored rather than answers to teachers' test questions provided and evaluated; in which teachers talk less than the usual two-thirds of the time and students talk correspondingly more. . . . Easy to imagine, but not easy to do. Observers have a hard time finding such discussions, and teachers sometimes have a hard time creating them even when they want to." (Cazden 2001, p. 54)

Despite the difficulty involved in orchestrating them, it is clear that robust discussions in the classroom are essential if students are to have opportunities to simultaneously engage in science practices and learn canonical science content. In the example of task B, the opportunity for students to present and compare their problem solutions resulted in a productive discussion that began when students noticed that Chesterton's warmest months occurred at the opposite time of year from Amber Lake and Bakersville. (This is a pattern that the class noticed after the group whose work is shown in fig. 0.7 shared their representation.) Once the class had noticed this general pattern, two questions emerged that led to further productive investigation: "Does spring always occur in March, April, and May?" and "Where are these cities located in the world?" By providing the students with task B and supporting the discussion that emerged when they shared their solutions, the teacher was able to help students learn that cities in the Northern and Southern Hemispheres experience the seasons at opposite times of the year at the same time that they were engaged in analysis and representation of data and communication of information.

Summary

Two crucial instructional challenges associated with the ambitious science education vision within the NGSS are (1) *designing and/or selecting instructional tasks* that provide opportunities for students to simultaneously engage with science practices and learn core concepts; and (2) *providing and managing opportunities* for students to talk productively with one another about their problem-solving approaches, solutions, models, etc. Such discursive interactions are at the heart of many of the targeted science practices in the NGSS (e.g., asking questions, constructing explanations, engaging in argument from evidence, and communicating information).

While establishing productive classroom talk remains a challenge for teachers (see Davis, Petish, and Smithey 2006), new tools and approaches have made this goal achievable, even for novice teachers. In this book, we describe how we have drawn upon groundbreaking work in mathematics education (Smith and Stein 2011) to implement an instructional model that enables teachers to learn how to notice and support student thinking through classroom discussion.

Laying the Groundwork: Setting Goals and Selecting Tasks

The knowledge, beliefs, and resources that teachers have all make a significant impact on their planning. For example, most teachers consult the available curriculum materials when setting learning goals and selecting tasks; and many teachers draw upon their understanding of their students' interests, academic strengths and weaknesses, social and cultural resources, etc., when planning lessons. The Next Generation Science Standards (NGSS), first published in 2013, are another factor that will now play a significant role in shaping instructional choices. In order to meet the goals of the NGSS, teachers will need to provide opportunities for students to engage in scientific practices (SPs) while exploring important phenomenological patterns and developing explanatory (conceptual) knowledge. The NGSS are based on a view stated in a report from the National Research Council (NRC) that "science is not just a body of knowledge that reflects current understanding of the world; it is also a set of practices used to establish, extend, and refine that knowledge. Both elements—knowledge and practice—are essential" (NRC 2012, p. 26).

In this chapter, we will discuss the general features of learning goals and tasks that are consistent with the vision of the NGSS, with the understanding that teachers will need to draw from a variety of resources to select and/or modify tasks to meet NGSS goals. Furthermore, while we acknowledge that teachers plan tasks to support a variety of activity structures (e.g., interactive lecture, collaborative group work, independent seatwork) within their classrooms, we focus here on tasks that teachers might use to engage learners in productive whole-class discussions. Later, in chapters 3 and 4, we will describe specifically how teachers might use the five practices to orchestrate such discussions and when, in a coherent arc of lessons, teachers might choose to conduct a Five Practices discussion (as described in chapter 6).

Identifying Instructional Goals

A teacher needs to have clear goals for what he or she is trying to accomplish in a lesson. It is important to develop goals in sufficient detail to support planning (e.g., selecting a task that is consistent with the desired outcomes) and instruction (e.g., responding to students as they engage in a lesson in order to help them advance toward the desired goals). Hiebert and colleagues argue that this level of specificity is critical to effective teaching:

> Without explicit learning goals, it is difficult to know what counts as evidence of students' learning, how students' learning can be linked to particular instructional activities, and how to revise instruction to facilitate students' learning more effectively. Formulating clear, explicit learning goals sets the stage for everything else. (2007, p. 51)

Figure 1.1 lists four potential goals for a series of sixth-grade lessons about Moon phases. Goals A and C are examples of **learning goals**—statements that describe what students will *know or understand* as a result of instruction. Goal A is extremely general, stating only that students will learn about the topic of Moon phases. It does not provide insight into the specific scientific ideas that students will develop. In contrast, goal C offers detail about the phenomenological patterns (the length of the Moon phase cycle, the order in which the phases appear, etc.) and explanatory knowledge (the Moon orbits the Earth; the relative positions of the Earth, Moon, and Sun account for the phase that is visible from Earth) that students should derive during the lessons.

Goal A:	Students will learn Moon phases.
Goal B:	Students will be able to describe Moon phases and explain why we (on Earth) see them.
Goal C:	Students will learn that we (on Earth) see different phases of the Moon throughout a one-month cycle. Following a New Moon, the Moon appears as a Waxing Crescent. Then we see the First Quarter, Waxing Gibbous, Full, Waning Gibbous, Third Quarter, and Waning Crescent Moons in successive order. The Moon orbits the Earth at rate of one complete revolution each month. The relative position of the Earth, Moon, and Sun determine how much of the illuminated portion of the Moon is visible from Earth. (For example, when the Moon is at the position in its orbit such that the Earth is directly between it and the Sun, people on Earth can see the entire illuminated face of the Moon. This phase is called the Full Moon.)
Goal D:	Students will use two- and three-dimensional models to demonstrate the relative positions of the Earth, Moon, and Sun during various Moon phases. For any particular arrangement of these celestial bodies, students will explain to their peers why the Moon would appear in a particular phase to observers on Earth.

Fig. 1.1. Four different goal statements for a series of sixth-grade lessons about Moon phases

Goals B and D provide information about what students will be *able to do* as a result of instruction. Thus, these are **performance goals**—statements that describe observable and measurable instructional outcomes. Like goal A, goal B is quite general. It states that students will be able to describe and explain Moon phases, but it leaves one wondering, "What *aspects* of Moon phases should students describe? What *specific patterns* should they account for? What is an *acceptable or sufficient explanation* for Moon phases? *How* will students explain Moon phases?" Goal C provides some of the specificity that is missing. It describes in detail the *specific patterns* that students should learn as well as *what information an explanation should include*. However, goal C does not address the issue of *how* students will offer their explanations. Goal D makes this clear by providing a specific description of what students will be *able to do* following the lesson. The specificity of learning goal C and performance goal D provides the teacher with clear targets that can guide the selection of tasks and the use of the five practices to support robust discussion during instruction.

Formulating clear learning and performance goals is an essential first step in lesson planning. Most K–12 teachers draw from curriculum materials when planning, and the format of such materials influences how teachers use them in significant ways. For example, some curriculum materials are provided in **scope and sequence format,** listing particular ideas or topics with which students should engage at various points in an academic year (see fig. 1.2, left side). Other curriculum materials specify certain tasks or instructional activities that teachers should implement (see fig. 1.2, right side). Regardless of the format of the curriculum materials provided, teachers should begin their planning by articulating learning and performance goals in sufficient detail to select and/or modify instructional tasks and to guide and support instruction and assessment.

Scope and Sequence	Lesson-Level Description
Unit 1: Force and Motion *A force is required to change an object's speed and/or direction.* **Unit 2: Patterns in the Sky** *The Earth is part of a larger Sun, Moon, Earth system. Objects in the sky have patterns that can be observed.* **Unit 3: The Water Cycle** *When liquid water disappears, it turns into a gas in the air. It can reappear as a liquid when cooled or as a solid when cooled further. Tiny droplets of water or ice in clouds fall to the ground as precipitation.*	**Unit 2: Patterns in the Sky** *Day 1* Read *The Big Dipper and You* by Edwin C. Krupp. Discuss the patterns that students have noticed in the sky. *Day 2* Introduce the major constellations visible in North America during each season. Use teacher's CD-ROM (chapter 3, section 1) to show images of major constellations. *Day 3* Planetarium field trip.

Fig. 1.2. Examples of curriculum resources for a third-grade science teacher. These topics and major ideas were adapted from the Pennsylvania Standards Aligned System, which is used statewide as a K–12 curriculum guide.

A third-grade teacher working from the Scope and Sequence shown in the left side of figure 1.2 might begin planning for unit 2 by asking: *What specific patterns should students notice?* The teacher might consult the NGSS and determine that students in grade 3 should know that the Sun appears to rise and set every twenty-four hours, and that throughout any particular day, it appears low on the eastern horizon, gradually climbs higher in the sky, and then sinks below the western horizon. These specific patterns are learning goals for unit 2. Knowing these learning goals, the teacher can then select tasks that will provide students with opportunities to notice these patterns (either through inquiry or more direct instruction).

Alternatively, if the teacher's curriculum is provided on a lesson level, as in the right side of figure 1.2, then he or she might begin by carefully reviewing each lesson task and asking, *What patterns should students notice as they participate in this task? What ideas or facts will students become familiar with?* After reading *The Big Dipper and You,* the teacher might conclude that the students will learn what the Big Dipper constellation looks like, as well as where and when it appears in the sky. Next, the teacher should formulate specific learning goals (e.g., the Big Dipper is a constellation that contains seven stars). The teacher may also want to consult the NGSS to determine whether other important learning goals should be addressed in the lesson. Having formulated these specific learning goals, the teacher is now able to make purposeful decisions about whether or how to modify a task and/or what types of scaffolding would assist students in their engagement of the task.

TRY THIS!

Select an instructional task provided within your curriculum. Identify the specific learning goals and performance goals described within the material. Develop detailed learning and/or performance goals if they are insufficiently described, or absent.

Assessing Tasks by Category and by Cognitive Demand

A variety of tasks might prompt productive discussions in science classrooms. We will focus here on three categories of tasks in particular: (1) *experimentation;* (2) *data representation, analysis, and interpretation;* and (3) *explanation.* Experimentation tasks involve students in designing, critiquing, and/or carrying out an experimental protocol. The second category of tasks involves students in representing, analyzing, and/or interpreting data. Jeremy's vacation task (fig. 0.4 on page 3), for example, fits into this category, as it involves students in representing data (constructing a graph) and interpreting patterns in the data. The last category of tasks includes those that involve students in providing explanations for patterns or phenomena. When used together, tasks in these three categories can provide opportunities for students to engage in all eight of the NGSS science practices (Achieve, Inc. 2013), an idea we discuss in greater detail in chapter 6.

One way of characterizing instructional tasks is to describe the level of cognitive demand required of students who engage in them (Doyle 1983; Stein, Grover, and Henningsen 1996). A task that requires students to *invest significant effort in making sense of the underlying science phenomena or concepts* is a high cognitive demand task. It is important to distinguish cognitive demand from other types of challenges associated with instructional tasks. For example, a task might be difficult for students because the text is complex (making it challenging for students to read the task with comprehension) or because the mathematics required to complete necessary computations is beyond their skills. A task that is challenging for reasons such as these is not necessarily cognitively demanding. For example, a teacher may ask students to read a section of text that is written at an advanced reading level beyond that of her students, and to answer a series of questions afterwards. If the questions merely ask students to copy information from the text, then the task, while challenging for struggling readers, is of low cognitive demand—there is no significant requirement for sense making related to the underlying content or phenomena. The challenge lies solely in the work of decoding and comprehending the text.

Teachers often make the mistake of assuming that students who struggle with textual or mathematical challenges are unable to successfully engage with cognitively demanding tasks. This is not the case. It is important for all students to have opportunities to learn science by participating in tasks that require them to think hard about the ideas and phenomena they are encountering. It is the responsibility of the teacher to select or design such cognitively demanding tasks while providing appropriate scaffolds to minimize the barriers that text or mathematical challenges might pose to participation.

Students' engagement in any of the three categories of science tasks described above— *experimentation; data representation, analysis, and interpretation;* and *explanation*—can be robust (involving a high level of cognitive demand) or perfunctory, depending upon the features of a particular task and the choices that the teacher makes during its enactment. In general, tasks that require students to make and justify choices about approaches or strategies involve high cognitive demand. In contrast, tasks that students can complete using an algorithmic approach, or those that require them to simply state an answer without providing a rationale, involve low cognitive demand. In the following sections, we describe some additional specific features of these three categories of tasks that contribute to the cognitive demand placed on students as they engage with them.

Experimentation Tasks

Experimentation tasks are ubiquitous in science classrooms. Usually, students follow a detailed protocol as they conduct their experiment. "Measuring Fast Plant Growth" (fig. 1.3a) is an example of this type of low-level experimentation task. Note that, first of all, the procedures that students must complete are described clearly and in detail; and, secondly, the task does not include an explicit connection to the underlying question that the experiment is designed to address. It is easy to imagine students following these procedures without having to engage in any sense making.

In contrast, "Choosing Materials for Umbrellas" (fig. 1.3b) is an experimentation task that involves a high level of cognitive demand. In this task, students are explicitly reminded of the purpose of the investigation (to determine how various materials perform when exposed to water). This encourages students to connect their hands-on activity with the underlying ideas. They are also told that they will have to design a protocol that "everyone has to understand." In other words, they will engage in the task with the anticipation of an audience for their work, one that will be a critical judge of it. Finally, this task involves students in making reasoned choices about the tools they will use in the experiment as well as how to use them. All of these features—explicit connection to purpose, an audience, and the need to make choices—contribute to the high cognitive demand of this task.

In addition to task features, the placement of an experimentation task in the overall instructional sequence also has an impact on its cognitive demand. In traditional science classrooms, students conduct experiments after the teacher has provided some didactic instruction about the underlying concept. In such a context, the experiment serves to provide confirming evidence of the concept already introduced. For example, a high school biology teacher might ask her students to read the text chapter about meiosis and sexual reproduction and then give a lecture in which she describes the mechanisms of independent assortment and fertilization. Students may subsequently engage in a virtual lab in which they are provided with parental organisms with known genotypes and prompted to predict the phenotypes of the offspring. After completing their predictions (which involves "running" the processes of independent assortment and fertilization, usually with a representational tool such as a Punnett square), students perform the indicated crosses and record data about the offspring. Finally they calculate the resulting phenotypic ratios (e.g., 3:1 dominant:recessive when both parents are heterozygous and one allele is completely dominant over the other). An experimentation task such as this one provides opportunities for students to *carry out an investigation* (NGSS Science Practice 3; see fig. 0.1 on page 1), *analyze data* (SP 4) by examining the phenotypic ratios of offspring, and *use mathematics* (SP 5). However, we would argue that this is a relatively low cognitive demand task because students are told exactly what to look for before beginning the experiment (ratios that are evidence of independent assortment and fertilization) in

Experimentation Tasks

Context
7th grade Biology

The teacher chose this task because she wanted the students to participate in data collection. Specifically, she wanted them to have an opportunity to make and record measurements over time. She chose Fastplants because she wanted students to learn that there is variation in "normal" growth in a population of plants, but that the general trend can be described by an s-shaped growth curve.

Measuring Fastplant Growth

1. Gently tie a piece of yarn around the base of each plant in your container. Be sure to use a different color yarn for each plant.

2. Prepare a length of measuring string:
   a. Cut a 24-inch segment of white string.
   b. Using a Sharpie marker, place a mark ½-1 inch from one end of the string.

3. Every two days measure the stem length of each plant:
   a. Place the black mark on your measuring string against the bottom of the plant stem. Make sure the black mark is right where the plant stem emerges from the soil.
   b. Gently run the string up the stem, stopping at the base of the highest flower cluster.
   c. Use your fingers to mark (by pinching off) the place where the stem ends.

4. Now use a meter stick to measure the length of the string from the black mark to the place where you have pinched.

5. Record each stem length measurement (in cm) in your data table:

Plant Height (cm)

	Plant 1 Green	Plant 2 Red	Plant 3 Blue	Plant 4 Yellow
Day 4	1.4	1.9	0.92	2.2
Day 6	3.2	3.8	2.4	4.6
Day 8	6.1	6.8	4.5	7.3

Context
3rd grade science

The teacher designed this task to provide students with an opportunity to gather data by performing and recording measurements. She also wanted students to participate in selecting measurement tools and designing the protocol so that they would learn about the importance of specificity and consistency in measurement. She embedded this task in a unit that focused on the properties and functions of materials so that students could also learn that some types of fabrics are better than others at repelling water.

[task by Elaine Lucas-Evans]

Choosing Materials for Umbrellas

The StayDri Company has asked our class to help them with product development. StayDri makes products that people use to protect things from getting wet. For example, one of their most popular products is a travel umbrella. The umbrella is a good product because it keeps rain off of people and it dries very fast after you bring it indoors.

StayDri wants us to test 8 different materials for a new and improved umbrella.

IMPORTANT FEATURES

The new umbrella needs to –

   a. Keep water off of people or things that are underneath it; and
   b. Dry quickly once it is out of the rain.

TESTING MATERIALS

We have the following tools available for testing the umbrella materials:

Water   Beaker   Markers
Water dropper   Food coloring   Ruler
Squirt bottle   Filter paper   Stopwatch

How will your group test each material to see how well it keeps water off of things?

Write out the steps of your test and draw pictures.
   Remember:
   • Everyone has to be able to understand how you will do your test.
   • Your test has to be fair. All of the materials have to be tested in the same way.

Fig. 1.3. Two examples of experimentation tasks: a low-level task (*a*), and a high-level task (*b*)

addition to precisely how to generate the data (which crosses to perform). Placing the experimentation task *before* the lesson in which the underlying causal mechanism is described can increase the cognitive demand for students. Moreover, experimentation that precedes explanation is consistent with the learning cycle, a framework we will discuss in greater detail in chapter 6.

Data Representation, Analysis, and Interpretation Tasks

Tasks that fall into this second category can also have features that add to or decrease the cognitive demand for students. The "Temperature Patterns" task (fig. 1.4a) is a low-level task because, while it does involve students in representing and analyzing data, it does not ask them to make any choices about how best to represent the data, nor does it prompt students to provide justification for their assertion about "where and when Jeremy should go on vacation." The task below it, "Environmental Factors Impacting Rate of Transpiration" (fig 1.4b), is a high-level data task. It requires students to examine data to identify patterns that are not immediately obvious in the table provided. In fact, students will have to use mathematical processes to transform the data (i.e., calculate the change in mass over time) in order to make patterns evident. Other features of this task that contribute to high cognitive demand include (*a*) students have to determine on their own the best way to represent the data that is relevant; and (*b*) students must prepare a written description of the patterns that will be convincing and understandable to the "Zoo Board." As we saw with "Choosing Materials for Umbrellas," the anticipation of an audience increases cognitive demand because it requires students to consider their representational and linguistic choices and to make explicit the data/claim connections and the justification for their approaches.

Explanation Tasks

Science students are often asked to provide explanations. The most significant differences between high- and low-level tasks of this type are, first, whether the student must provide a rationale for the explanation (e.g., support the claims he or she makes with evidence); and, second, whether the student constructs the explanation (e.g., it is the result of meaning making) or whether the student is simply repeating an explanation that he or she has been told previously. For example, during a series of lessons about Moon phases, a teacher might explain that the reason we see the Moon changing phase is that it revolves around the Earth each month, and as it does so, different parts of the illuminated side of the Moon are visible from Earth. Later, the teacher might ask her students, "Explain why we see Moon phases." Students who remember the teacher's explanation can simply repeat or rephrase it in answer to her prompt. Thus, the explanatory task places low cognitive demand on these students. In contrast, "The Frog Problem in Bakersville Park" (fig. 1.5) is an explanatory task that places high cognitive demand on students. In this task, students are asked to explain what is causing the frog deformities in the park's lakes. To construct this explanation, students are prompted to "use the data . . . to support or challenge one of the hypotheses." They have multiple options for how to approach the problem (i.e., they can draw from the different data sources, transform or represent the data as needed, etc.). Similar to the task "Environmental Factors Impacting Rate of Transpiration," the Frog Problem task is also made more challenging because the data with which students are asked to reason are complex (e.g., units are not consistent and therefore students cannot simply compare quantities). Moreover, the task is challenging for students because it requires them to determine the most effective way to transform and represent data in order to persuade their peers of the validity of their argument.

Data Representation, Analysis, and Interpretation Tasks

Context	Temperature Patterns
(a) 6th grade Earth Science The teacher selected this task in order to give his students an opportunity to create and read bar graphs.	Jeremy is planning ahead for his 2015 vacation. He has decided that he'd like to travel to a place where he can enjoy outdoor camping, hiking, and fishing with his Labrador retriever, Sadie. Jeremy's tent is rated for temperatures above freezing (32 °F). Sadie prefers not to be too active when the temperature is over 70°F. Create a bar graph that shows the average monthly high and low temperatures in each city. Identify where and when Jeremy should go on vacation. (See data for Task A, Fig. 0.2).

Context	Environmental Factors Impacting Rate of Transpiration
(b) 9th grade Biology The teacher designed this task to provide students with an opportunity to make choices about how to transform data (e.g. calculate the change in mass over time) and represent it in order to show trends that would enable them to answer a specific question. She embedded the task in the context of a unit on respiration and thus highlighted key Learning Goals related to the role of water in plant transpiration. [task by Helen Snodgrass, KSTF Fellow]	Dear scientists of Prep HS, We are writing you as fellow scientists in need of some help. At the zoo, our expertise is mainly in the area of animals and we currently have a question about our plants that we hope you can help with. In different areas of the zoo, plants experience variable growth conditions. Some areas are more humid or shadier than others, etc. We need to develop a plan to provide the correct amount of water to our plants. That watering plan has to take into consideration the rate of transpiration of the plants under different conditions. Our grounds crew has gathered some data about the plants over a 5-day period during which the plants received no water. We would like you to use this data to develop a report about how different environmental growth conditions impact rate of transpiration. Once we receive your report, we can develop a watering plan that will enable us to keep our zoo habitats thriving! We need to present this data to the Zoo Board at its next meeting. **Please look over the data for any patterns you see and create a graphical representation so that we can show the board members what patterns you have identified. Also, it will be very important to have some written description of what you found out so that our Zoo Board members will be convinced that our watering plan is grounded in good science.** Thank you for your help. We are looking forward to hearing from you. Deborah Smith Director of the Zoo

Variable Condition	Standard Growth Conditions	Mass (g) Day 1	Mass (g) Day 2	Mass (g) Day 3	Mass (g) Day 4	Mass (g) Day 5
---------	64-87°F 75% humidity 8-10 hours of sunlight/day 10 mph winds	16.0	13.2	11.0	9.9	9.0
90% humidity	64-87°F 8-10 hours of sunlight/day 10 mph winds	17.0	16.8	16.6	16.4	15.3
2 hrs of sunlight	64-87°F 75% humidity 10 mph winds	12.9	12.5	11.9	11.4	11.1
40 mph winds	64-87°F 75% humidity 8-10 hours of sunlight/day	16.3	12.6	9.8	7.7	5.1

Fig. 1.4. Two examples of data representation, analysis, and interpretation tasks: a low-level task (*a*), and a high-level task below it (*b*)

Explanation Task

Context	The Frog Problem in Bakersville Park
5th grade science The teacher designed this task to provide students with an opportunity to draw on data to make and defend claims. She embedded the task in a unit about ecosystems, anticipating that students would draw upon their understanding of how organisms interact with and are dependent upon living and non-living factors in their environments. She wanted them to build on this knowledge to learn that parasites (or other pollutants in an ecosystem) can be particularly problematic for organisms that are exposed during early stages of development. After the students presented and discussed their claims, she took time to emphasize this new Learning Goal before closing the lesson.	Visitors to Bakersville Park have been noticing some strange looking frogs in and around some of the ponds!  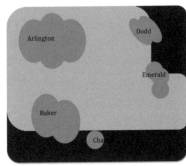 Around Baker, Charles, and Emerald ponds, they have been seeing frogs with too few or too many legs! None of the deformed frogs have been spotted around Arlington or Dodd ponds, though. Local scientists are wondering: **what is causing these strange deformities?** They have two hypotheses: 1. There is some kind of chemical pollution in Baker, Charles, and Emerald ponds that is causing the frogs to be deformed. 2. There is a disease-causing organism (a bacterium or parasite) in these ponds that is causing the deformities. **Use the data that the scientists have collected to support or challenge one of the hypotheses.**

DATA
Concentration of Chemical Pollutants in Bakersville Park Ponds

	Fertilizer Pollution Level (ppm)	Pesticide Pollution Level (ppm)
Arlington	37	11
Baker	43	17
Charles	34	8
Dodd	41	22
Emerald	28	21

ppm = parts per million

Presence of Tremadode Larvae in Frogs

	number of frogs that were NOT infected	*number of frogs that were infected*	**Percentage** of Frogs Infected by Trematodes
Arlington	24	1	4
Baker	16	9	36
Charles	14	11	44
Dodd	23	2	8
Emerald	15	10	40

Fig. 1.5. An example of an explanation task with high cognitive demand

The Teacher's Role

As noted in the outset of this chapter, we are particularly interested in instructional tasks that (*a*) provide students with opportunities to learn key science ideas while also engaging in important disciplinary practices; and (*b*) are robust enough to support a productive whole-class discussion following students' engagement in the tasks. By "productive whole-class discussion" we mean one in which students share ideas, focus on meaning making, and develop new or richer understandings of key concepts. To support such discussion, the teacher must ensure that the following conditions are met:

1. The task places **high cognitive demand** on students, and the teacher's instruction serves to maintain, rather than remove or minimize, that demand.

2. Students are able to engage in the task in **multiple ways** that are productive (i.e., that contribute to the achievement of the learning goals). This is important because the whole-class discussion provides an opportunity for students to share their ideas and to listen critically to others. If all students have the same ideas or take the same approach to a task, they have no incentive to attend closely to one another, and no opportunity to make comparisons or connections. Moreover, providing a task in which students can engage in different ways helps to promote equity in the classroom, enabling all students to draw upon their particular experiences and cognitive resources to participate in the learning context.

3. Students **produce artifacts** while engaged in the task. Artifacts may include written text or drawings that serve multiple purposes. First, they function as a tool to support the students' thinking (and their communication about their thinking when working with others) during the task. Second, they provide the teacher with important information about the students' ideas and with opportunities to ask questions that can help to redirect or push student thinking. Finally, the artifacts serve as a tool to focus and support the subsequent whole-class discussion. They capture key elements of students' work and therefore function to center the discussion on those features.

Teachers include many different types of activity structures in their classrooms (e.g., lecture, seatwork, collaborative group activities). Some activity structures are more useful than others as precursors to whole-group discussion. For example, collaborative group work is an activity in which students are able to generate a variety of ideas or approaches related to a task and to produce artifacts that capture those ideas. In contrast, lecture and note-taking are activities that do not meet the conditions described above for supporting productive whole-class discussions. Figure 1.6 depicts many common activity structures used by science teachers. It indicates that those involving small groups of students working collaboratively are most appropriate for setting up a Five Practices discussion.

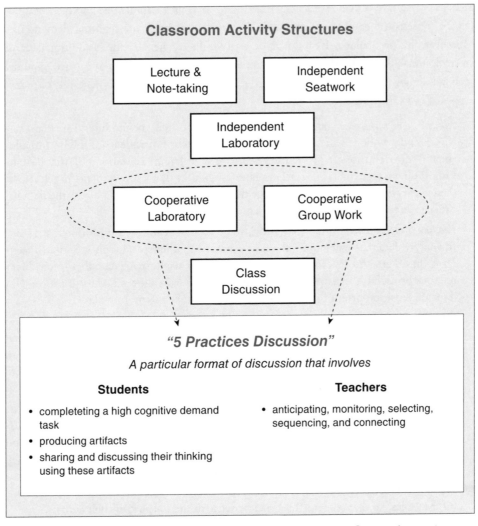

Fig. 1.6. An assortment of common classroom activity structures. Cooperative group activities (including laboratory tasks) are the ones most likely to support productive whole-class discussion.

Modifying Tasks

Science teachers select instructional tasks from curriculum materials such as science kits and text-books, as well as from a variety of online resources. Often, teachers find that the tasks that are readily available place low cognitive demand on students (similar to the tasks shown in figs. 1.3a and 1.4a). In such situations, teachers can make specific modifications to tasks, or strategic choices about the enactment of tasks, that will serve to increase their cognitive demand. For example, a teacher whose curriculum materials include "Measuring Fast Plant Growth" (fig. 1.3a) might decide to alter the task so that students are responsible for developing the measurement protocol them-selves, such as shown in the task "Studying Fast Plant Growth" (fig. 1.7). By providing students with a variety of tools and asking them to design their own measurement protocols, the modified task requires students to make meaning of their actions rather than simply follow rote directions.

The teacher's decision to provide students with time to share and critique one another's designs (and to develop a consensus measurement protocol) also serves to increase the cognitive demand of the task. Moreover, this particular task modification enables the teacher to address additional learning goals in the lesson—goals related to students' understanding of key features of experimental design. Some general design strategies that teachers can use to increase the cognitive demands of many different types of tasks include:

1. *Eliminate or minimize prescriptive directions.* For example, the modified Fast Plant task (fig. 1.7) does not provide a highly detailed set of steps for students to follow, but allows them to develop those steps themselves. Or, as with Jeremy's vacation task (fig. 0.4) and the Frog Problem task (fig. 1.5), teachers can design tasks that allow students to select which data to represent, how to transform the data, and/or how to best represent the data in order to support a particular claim or conclusion.

2. *Provide complex data.* Rather than providing data that is already transformed, ask students to analyze data that will require them to use some mathematical tools in order to see patterns (figs. 1.4b and 1.5, for example). Teachers can also provide data that is not directly relevant or useful for answering the questions posed, and allow students to reason which data are most important for supporting the claims they intend to make.

3. *Give students an audience.* Providing an opportunity for students to present their work and to critique that of peers increases the cognitive demand of tasks. This implementation approach forces students to consider the linguistic and representational choices they make to express their ideas, and it requires them to make connections across ideas while actively listening to peers.

4. *Re-sequence tasks.* As noted earlier, traditional science instruction often involves a didactic lesson in which students receive information about causal mechanisms or concepts followed by a laboratory exercise in which they generate empirical evidence that supports these concepts. A teacher can provide more opportunities for students to engage in sense making by placing the exploratory laboratory first in the sequence of lessons. Such exploratory laboratory exercises must still be firmly grounded in a question (see figs. 1.3a and 1.7 for examples) so that students have a clear sense of the purpose of their activity.

TRY THIS!

Choose a task from one of the three categories described in this chapter. Identify (1) the existing features of the task that would place high cognitive demand on students, and (2) specific modifications you might make to the task in order to increase its cognitive demand.

Maintaining Cognitive Demand during Task Enactment

Task selection and design are crucial to ensuring that students have opportunities to engage in high cognitive demand work. However, a teacher's choices during the enactment of a task also have a significant impact on the cognitive demand that students experience. Moreover, researchers in the field of mathematics have shown a positive relationship between teachers' ability to maintain high cognitive demand of tasks during enactment and student learning (Stein and Lane 1996; Hiebert and Stigler 2004; Boaler and Staples 2008).

Studying Fastplant Growth

We know that individual humans vary quite a lot from one another — we are different heights and weights; we have different skin, hair, and eye color; the thickness of our hair varies, etc.

Is there variation in populations of other types of organisms?

- **Would we see variation in a population of plants?**
- **What kind of variation would we see?**
- **How would we measure and describe that variation?**

Over the next few weeks you will be investigating variation in a population of plants called Wisconsin Fastplants. We are going to track **changes in stem length** as the plants grow.

Today we will decide how we are going to measure stem length in Fastplants.

SMALL GROUPS
[20 minutes]

1. Obtain a Fastplant from under the grow lights.
2. Select from the available tools:

Measuring tape	Markers	Lego blocks
Bamboo skewers	Colored tape	Pipe cleaners
String	Meter stick	
Scissors	Ruler	

3. Determine how you will use the tool/s you've chosen to measure Fastplant stem length.

4. Write our your measurement protocol in enough detail so that others will be able to use the protocol in a reliable way (i.e. everyone needs to be able to use it exactly the same way).

   Include pictures to help others understand your measurement protocol.

WHOLE CLASS
[20 minutes]
- We will share our protocols with the class and determine whether there are any details missing.
- We will agree on one way of measuring our plants throughout this investigation.

Fig. 1.7. In this modified version of the task "Measuring Fast Plant Growth" (fig. 1.3a), students are given clear instructions to connect the data collection task to an underlying question ("How would we measure and describe that variation?"). They also have choices about what tools to use and how to use them to obtain measurement data as well as the opportunity to share and critique approaches with peers. These modifications serve to increase the cognitive demand of the task.

The table in figure 1.8 summarizes some of the key features and teacher actions that contribute to low and high cognitive demand enactments of three types of tasks in science. For example, teachers who provide opportunities for students to share and critique will help to maintain the high cognitive demand of explanatory tasks. Teachers' actions, it should be noted, often serve to *lower* the cognitive demand (even for robust tasks), and it is therefore crucial that teachers are purposeful about their actions in order to support students' engagement in challenging tasks (Stein, Grover, and Henningsen 1996). In chapters 2 through 5, we will present a more detailed look at how the Five Practices framework and its deliberate strategies to elicit and support student talk can help teachers to ensure students' productive engagement in high cognitive demand tasks.

| | **Low Cognitive Demand** | | **High Cognitive Demand** | |
	Tasks	*Teacher Actions*	*Tasks*	*Teacher Actions*
Experimentation	Students— • follow a **highly specified** procedure. • do not make choices about what data to collect or how to collect it. • are not engaged in being critical about the data collection procedure.	The teacher— • does not help students understand that data collection is occurring in the service of answering a question. • introduces the experiment after she/he has already provided didactic information on the underlying concepts.	Students— • must **make decisions** about **what** data to collect and/or **how** to collect it. • **compare/contrast or critique** experimental protocols, considering issues such as reliability and "fit" between data gathered and the underlying question driving the experiment.	The teacher— • ensures that students understand how their data collection must help them achieve the goal of answering a particular question.
Data Representation, Analysis, and Interpretation	Students— • **follow specific instructions** about how to transform (e.g., calculate the mean temperature) and/or represent data (e.g., draw a bar graph). • **answer specific questions** about the data (e.g., *In which city is the average monthly temperature highest?*).	The teacher— • accepts only very specific representation types or strategies. (i.e., multiple solutions or strategies are not possible). • does not press for students to justify their answers using the data representations.	Students— • **seek to describe general** (e.g., the S-shaped growth curve of Fast Plants) and specific (e.g., trematode infection is 4–5 times higher in Charles, Emerald, and Baker ponds than in other ponds) **patterns** that are evident in the data. • **select** what data to represent and/or **how** to represent it. • **compare/contrast** various representations, considering issues such as the ease with which various patterns or relationships can be visualized.	The teacher— • provides opportunities for students to share and discuss a variety of data representations. • requires students to provide a rationale for the choices they have made related to transforming or representing data. • requires students to identify specific data or elements of data representations that provide evidence for the patterns/trends they've identified.

	Low Cognitive Demand		High Cognitive Demand	
	Tasks	**Teacher Actions**	**Tasks**	**Teacher Actions**
Explanation	Students— • **provide explanations without justification** or specific connection to data. • **repeat factual knowledge** previously learned.	The teacher— • requests discrete answers to questions without justification (e.g., *What causes a solar eclipse?* [answer] *The Moon blocking the Sun.*)	Students— • **provide explanations** with justification. • are engaged in **developing new explanatory knowledge.** • are critical of the explanations offered by others, requesting clarification and supporting evidence when appropriate. • **draw upon a variety of representational tools** (e.g., diagrams, tables, simulations) to communicate with peers.	The teacher— • presses students to provide explanations and to justify their assertions. • provides opportunities for students to share and critique one another's explanations. • encourages students to use a variety of tools to communicate.

Fig. 1.8. The task features and teacher actions that contribute to low or high cognitive demand

CHAPTER 2

Introducing the Five Practices Model: Contrasting the Practices of Two Teachers

In this book's introduction, we indicated that while robust classroom discussions are essential if students are to simultaneously engage in science practices and learn canonical science content, they are difficult to orchestrate. Why is it so challenging for teachers to orchestrate productive discussions? Research tells us that students learn when they are encouraged to become the authors of their own ideas *and* when they are held accountable for reasoning about and understanding key ideas (Engle and Conant 2002). In practice, doing both of these at the same time is very difficult. By their nature, high cognitive demand tasks that engage students in experimentation; data representation, analysis, and interpretation; and explanation (as discussed in chapter 1) do not lead to cookie-cutter products. Rather, teachers can and should expect to see varied (incorporating both correct and incorrect ideas or strategies) responses to a task during the discussion phase of the lesson. In theory, this variety is a good thing because students are "authoring" (or constructing) their own ways of making sense of the situations presented.

The challenge rests in the fact that teachers must also align the many disparate ideas and approaches that students generate in response to high cognitive demand tasks with the learning goals of the lesson. It is the teacher's responsibility to move students collectively toward, and hold them accountable for, the development of a set of ideas and processes that are central to the discipline—those that are widely accepted as worthwhile and important in science as well as necessary for students' future learning of science in school. If the teacher fails to do this, the balance tips too far toward student authority, and classroom discussions become unmoored from accepted disciplinary understandings.

The key is to maintain the right balance. Too much focus on accountability can undermine students' authority and sense making and, unwittingly, encourage increased reliance on teacher direction. Students quickly get the message—often from subtle cues—that engaging in science practices means using only those strategies that have been validated by the teacher or textbook; correspondingly, they learn not to use or trust their own reasoning. Too much focus on student authorship, on the other hand, leads to classroom discussions that are free-for-alls.

Successful or Superficial? Discussion in Kelly Davis's Classroom

In short, the teacher's role in discussions is critical. Without expert guidance, discussions in science classrooms can easily devolve into the teacher taking over the lesson and simply telling students what to do and how to do it, or into the students presenting an unconnected series of show-and-tell presentations, all treated equally and illuminating little about the canonical science ideas that are the goal of the lesson. Consider, for example, the following vignette, featuring a seventh-grade teacher, Kelly Davis.

ACTIVE ENGAGEMENT 2.1

As you read the Case of Kelly Davis, identify the instances in it of student authorship of ideas and approaches, as well as any instances of holding students accountable to the discipline.

Growing Fast Plants: The Case of Kelly Davis

Ms. Davis wanted her grade 7 students to have an authentic experience collecting and representing data. To achieve this, she had students gather data on the growth of Wisconsin Fast Plants over an eleven-day period. (More information on these plants, which were developed by a program at the University of Wisconsin–Madison, is available at www.fastplants.com.) Each group of students tracked the growth of six Fast Plants by measuring their height every few days. They gathered data beginning on the day the plants were ten days old and ending when they were twenty-one days old.

Following data collection, Ms. Davis asked the students to create a representation of their data on poster paper that would enable them to answer the following question: "How tall is a typical Fast Plant on a certain day in its life cycle?"

Ms. Davis told her students that they could represent their data any way they wanted. She also told them they could either use their raw data (the actual values they recorded) or else transform the data in some way. She emphasized that, however they chose to represent their data, they needed to be able to explain what values they plotted, how they got those values, and why their representation helps to answer the question "How tall is a typical Fast Plant on a certain day in its life cycle?"

As students worked in groups, Ms. Davis walked around the room making sure that students were on task and making progress on the activity. She was pleased to see that students were using many different approaches to represent their data—different formats (bar graphs and line graphs) and different mathematical measures of central tendency (mean, median).

She noticed that students in two of the groups were having some difficulty accurately representing their data. Students in group 3 decided to calculate the

mean plant height at each data collection point but made mathematical errors in their calculations, and students in group 4 used inconsistently spaced units when constructing their line graphs. Ms. Davis told both these groups that
30 they had an error in their poster and that rather than presenting, they needed to listen carefully to the presentations from the other groups and should try to determine what they needed to fix. Ms. Davis was not too concerned about these errors, however, since she felt that once several correct representations were shared, students in groups 3 and 4 would see what they did wrong and learn
35 new strategies for creating correct graphs in the future.

When most of the students were finished, Ms. Davis called the class together to discuss their work. She began the discussion by asking for volunteers to share their posters, being careful to avoid calling on the students with incorrect graphs. Over the course of the next fifteen minutes, groups 1, 5, 7, 9, 2, 6,
40 and 8 volunteered to present their representations to the class (see fig. 2.1). They described *typical* by using conventional methods such as the mean, median, and range, as well as less common approaches such as looking at the average of two middle plants each day.

During each presentation, Ms. Davis made sure to ask each presenter to
45 explain what values they had plotted, how they got the values, and the height of a typical plant. She also made sure that after each presentation she asked the class if they had any questions for the group presenting. A few students from group 5 (who just picked one of their six lines as typical without finding an average) asked questions about how other groups had calculated the mean
50 (group 8) and the median (group 2).

She concluded the class by telling students that the question "How tall is a typical Fast Plant?" could be answered in many different ways and that now, when they encountered a question like this, they could pick the way they liked best because all of these approaches gave them an answer.

Analyzing the Case of Kelly Davis

Some would consider Ms. Davis's lesson exemplary. Indeed, Ms. Davis did many things well, including allowing students to construct their own way of representing data and stressing the importance of students' being able to explain how their representation helped determine the height of a typical Fast Plant. Students worked in small groups and publicly shared their representations with their peers. They also had opportunities to engage in science practices related to data representation (SP 4) and communication of information to others (SP 8). All in all, students in Ms. Davis's class had the opportunity to become the "authors" of their own knowledge.

However, a more critical eye might note that the string of poster presentations did not build toward important ideas in science. While students did engage in disciplinary practices (e.g., analyzing and interpreting data), the lesson was narrowly focused on representing the data and did

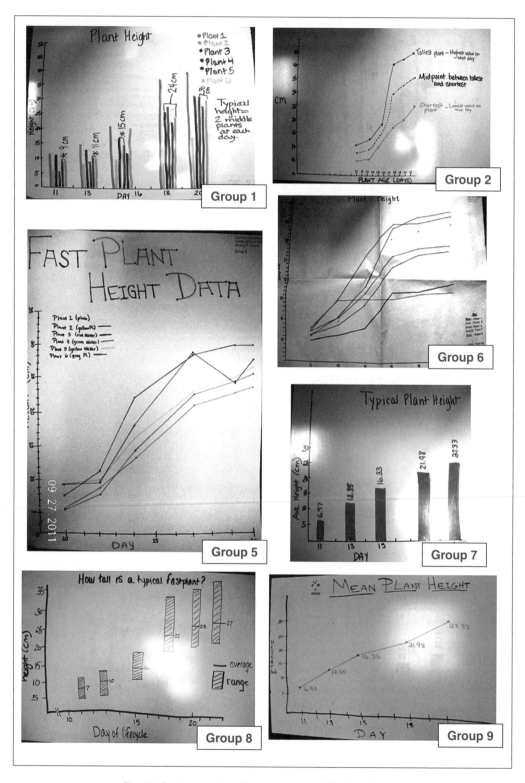

Fig. 2.1. Posters produced by students in Kelly Davis's class

not include support for students to learn key science ideas related to "typical" growth patterns. Such ideas could include how populations of organisms are diverse, so a range, rather than a single value, is generally better able to capture "typical"; or how the Fast Plant growth curve is almost always S-shaped due to the ways in which the plant is utilizing energy resources at different points in its life cycle. In addition, although Ms. Davis observed students as they worked, she did not appear to use this time to assess what they understood about their representations or to select particular students' work to feature in the whole-class discussion. Furthermore, she gathered no information regarding whether groups 3 and 4, who did not accurately represent their data, were helped by the student presentations. Had they diagnosed the faulty reasoning in their approaches?

In fact, we argue that much of the discussion in Ms. Davis's classroom was show-and-tell, in which students with accurate representations each took turns sharing their posters. The teacher did minimal filtering of the ideas that each poster helped to illustrate, nor did she make any attempt to highlight those ideas. The teacher also did not draw connections among different representations or tie them to important disciplinary methods or ideas. She gave no attention to weighing which representations might be most useful, efficient, accurate, and so on, in describing the height of a typical plant. All were treated as equally good.

In short, providing students with a high-level task and then conducting show-and-tell discussions cannot be counted on to move an entire class forward in their understanding of how to engage in science practices or their understanding of the key science ideas underlying the task. By arbitrarily sequencing one group presentation after another with limited teacher/student commentary and providing no help in drawing connections among methods or tying those methods to shared disciplinary concepts, Ms. Davis gave her students no motivation to attend to or understand their classmates' methods. Indeed, this kind of practice has been criticized for creating classroom environments in which nearly complete control of the agenda is relinquished to students. Some teachers misperceive the appeal to honor students' thinking and reasoning as a call for a complete moratorium on teacher shaping of the quality of students' thinking. As a result of the lack of guidance with respect to what teachers *could* do to encourage rigorous thinking and reasoning, many teachers are left feeling that they should avoid telling students anything.

In sum, Ms. Davis did little to encourage accountability to the discipline of science. How could Ms. Davis have more firmly supported student accountability without undermining student authority? The single most important thing that she could have done would be to set a clear goal for which crosscutting concepts and core ideas she wanted students to learn from the lesson. Without a learning goal in mind that went beyond "representing data," the various graphs presented were only discussed at a surface level. Key ideas in the discipline that could have been explored were not on the teacher's radar. If, however, she had targeted the learning goal that *typical growth in Fast Plants—and many populations of organisms—is described by range and shape*, she might have monitored students' work with this in mind, noticing whose work illustrated the range and shape particularly well. This assessment of student work would have allowed her to be more deliberate about which groups she selected to present during the discussion phase. With an array of purposefully selected representations presented, Ms. Davis would then have been in a position to steer the discussion toward a more satisfying conclusion.

The Case of Kelly Davis illustrates the need for guidance in shaping classroom discussions and maximizing their potential to extend students' thinking and connect it to important science concepts and core ideas. In the next section we offer this guidance by elaborating the Five Practices model, a practical method for orchestrating and managing productive classroom discussions.

The Five Practices Model

We think of the Five Practices model as skillful improvisation. The practices that we have identified are meant to make student-centered instruction more manageable by moderating the degree of improvisation required by the teacher during a discussion. Instead of focusing on in-the-moment responses to student contributions, the practices emphasize the importance of planning. Through planning, teachers can anticipate likely student contributions, prepare responses that they might make to those contributions, and decide how to structure students' presentations to further their learning agenda during a lesson. We turn now to an explication of the five practices.

The five practices were designed to help teachers use students' responses to advance the scientific understanding of the class as a whole during task-based discussions. They provide teachers with some control over what is likely to happen in the discussion as well as more time to make instructional decisions by shifting some of the decision making to the planning phase of the lesson. The five practices are—

1. *anticipating* how students are likely to respond to a task;
2. *monitoring* what students actually do as they work on the task in pairs or small groups;
3. *selecting* particular students to present their work during the whole-class discussion;
4. *sequencing* the student work or products that will be displayed in a specific order; and
5. *connecting* different students' responses and connecting the responses to key scientific ideas.

Each of these practices are described in more detail in the following sections, which illustrate them by identifying what Ms. Davis *could have done* in the Growing Fast Plants lesson (presented earlier in this chapter) to move student thinking more skillfully toward the goal of understanding that typical growth in Fast Plants—and in many populations of organisms—is described by range and shape.

Anticipating

The first practice is to actively envision how students might approach the instructional tasks or activities on which they will work. This involves much more than simply evaluating whether a task is at the right level of difficulty or of sufficient interest to students, and it goes beyond considering whether or not they are likely to get the "right answer." Anticipating students' responses involves developing considered expectations about how they might interpret a problem, the array of strategies—both correct and incorrect—that they might use to tackle it, and how those strategies and interpretations might relate to the concepts, representations, procedures, and practices that the teacher would like his or her students to learn.

Anticipating requires that teachers engage in the task or activity themselves, and consider different ways it could be approached or interpreted. Sometimes teachers find it helpful to expand what they might be able to think of individually by working on the task with colleagues, reviewing responses to the task that might be available (e.g., work produced by students in the previous year, or sample responses that are published along with tasks in supplementary materials), and consulting research on student learning of the scientific ideas embedded in the task. One resource for educators is the National Science Digital Library (NSDL), located at http://nsdl.org. NSDL provides information for educators and researchers on learning in science, technology, engineering, and mathematics (STEM). Their Science Literacy Maps are available to educators as a resource on specific math and science concepts. These maps clearly indicate specific benchmarks correlated to National Science Education Standards (NRC 1996), make connections between concepts across grade levels, and detail how those concepts build on one another. The maps also provide teaching and learning resources as well as document research-based student misconceptions for certain concepts.

Prior to the lesson, Ms. Davis needed to consider the various representations that students might use in answering the question she posed and give some thought to which ones would best highlight the two key features of determining what is typical in a population—range and shape. In addition, she needed to identify common errors students make. When they create graphs, students sometimes use a scale that is not consistent, or they graph dependent and independent variables on the wrong axes. In calculating measures of central tendency, students may make a number of errors. When finding the mean, they may add wrong or divide by the wrong number; they may forget to put data in order before finding the middle value when determining the mode; and they may think there can be only one median. By knowing in advance the errors students are likely to make, Ms. Davis would have been prepared to ask specific questions to help get students back on track rather than hoping that any misunderstandings would be alleviated by simply seeing correct approaches.

Monitoring

Monitoring the responses students produce involves paying close attention to their thinking and strategies as they work on the task. Teachers generally do this by circulating around the classroom while students work either individually or in small groups. Carefully attending to what students do and say as they work makes it possible for teachers to use their observations to decide on what and whom to focus during the discussion that follows (Lampert 2001).

One way to facilitate the monitoring process is for the teacher, before beginning the lesson, to create a list of anticipated student solutions or ideas that will help in accomplishing the lesson goals. The list, such as the one shown in the first two columns of the chart in figure 2.2 for the Growing Fast Plants task, can help the teacher keep track of which students or groups produced or brought up particular solutions or ideas that he or she wants to capture during the whole-group discussion. The "Other" category in the second column provides the teacher with the opportunity to capture ideas that he or she had *not* anticipated.

As discussed earlier in the chapter, Ms. Davis's lesson provided limited, if any, evidence of active monitoring. Although Ms. Davis knew which groups produced correct graphs and that a range of representations had been used, the fact that all seven groups with accurate graphs presented, regardless of the fact that they did not all uniquely contribute something to the discussion, suggests she had not considered the specific learning potential available in any of the responses. What Ms. Davis could have done while students worked on the task is shown in the right-hand column in the chart in figure 2.2.

Typical	Representation Type and Group Number	Notes and Order
Not specified	___ Bar √ Line—*group 5* ___ Box-and-whisker ___ Other	*Plotted data for all plants.*
Mean	√ Bar—*group 7* √ Line—*group 6; group 9* ___ Box-and-whisker √ Other—*group 8*	*Groups 7 (#2) & 9—plotted only mean.* *Group 6—plotted all individual plants and mean. (#1)* *Group 8—showed range and mean in a modified bar graph. (#3)*
Median	√ Bar—*group 1* ___ Line ___ Box-and-whisker ___ Other	*Identified "typical" as the mean value of the two plants "in the middle" of the data set.*
Mode	___ Bar ___ Line ___ Box-and-whisker ___ Other	
Range	___ Bar √ Line: *group 2* ___ Box-and-whisker √ Other: *group 8*	*Group 2—plotted highest, lowest, and middle value.* *Group 8—showed range in the bars. Also showed mean. (#3)*
Shape	___ Bar √ Line—*groups 2, 5, 6* ___ Box-and-whisker ___ Other	*Can see an S shape best in these. No group indicated this shape is "typical." (#1)*
ERRORS	√ Calculation—*group 3 (mean)* √ Graphing—*group 4 (inconsistent units across x-axis)*	

Fig. 2.2. A sample chart for monitoring students' work on the Growing Fast Plants task

Monitoring involves more than just watching and listening to students. During this time, the teacher must also ask questions that will make students' thinking visible, help them to clarify their thinking, ensure that members of the group are all engaged in the activity, and press them to consider aspects of the task to which they need to attend. Many of these questions can be planned in advance of the lesson, on the basis of the anticipated responses. For example, if Ms. Davis had anticipated that students would use a line graph (groups 2, 5, and 6; see fig. 2.1), then she might have been prepared to question the students regarding what they noticed about the shape of the graphs and what they thought the shape meant. Questioning a student or group of students while they are exploring the task provides them with the opportunity to refine or revise their thinking prior to whole-group discussion, and it provides the teacher with insights regarding what the students understand about the task and the ideas embedded in it.

Selecting

Having monitored the available student responses, the teacher can then select particular students to share their work with the rest of the class in order to get particular ideas on the table, thus giving the teacher more control over the discussion (Lampert 2001). The selection of particular students and their approaches or ideas is guided by the goals for the lesson and the teacher's assessment of how each response will contribute to those goals. Thus, the teacher selects certain students to present because of the concepts or core ideas in their responses.

A typical way to accomplish "selection" is to call on specific students (or groups of students) to present their work as the discussion proceeds. Alternatively, the teacher may let students know before the discussion that they will be presenting their work. In a hybrid variety, a teacher might ask for volunteers but then select a particular student that he or she knows is one of several who have a particularly useful idea to share with the class. By calling for volunteers but then strategically selecting from among them, the teacher signals appreciation for students' spontaneous contributions, while at the same time keeping control of the ideas that are publicly presented.

Returning to the Case of Kelly Davis, if we look at the solutions that were shared, we note that groups 5 and 6 had similar line graphs and groups 7 and 9 each graphed the mean height of the plants at each time point (although they used different types of graphs). Therefore, Ms. Davis might have considered only sharing one graph from each of these sets and spending more time discussing them.

Sequencing

Having selected particular students to present, the teacher can then make decisions regarding how to sequence the presentations. By making purposeful choices about the order in which students' work is shared, teachers can maximize the chances of achieving their goals for the discussion. For example, the teacher might want to have the response produced by the majority of students presented before those developed by only a few students in order to validate the work that the majority did and make the beginning of the discussion accessible to as many students as possible. Alternatively, the teacher might want to begin with a strategy that is more concrete (using drawings or concrete materials) and move to strategies that are more abstract. This approach—moving from concrete to abstract—serves to validate less sophisticated approaches and allows for connections between approaches. If a common misconception underlies a response offered by several students, the teacher might address it first so that the class can clear up that misunderstanding and work on developing more successful ways of tackling the task. Finally, the teacher might want to have related or contrasting ideas presented right after one another in order to make it easier for the class to compare them. Again, during planning the teacher can consider possible ways of sequencing anticipated responses to highlight the ideas that are key to the lesson. Unanticipated responses can then be fit into the sequence as the teacher makes final decisions about what is going to be presented.

More research needs to be done to compare the value of different sequencing methods, but we want to emphasize here that deliberate sequences can be used to advance particular goals for a lesson. Returning to the Case of Kelly Davis, we point out one sequence that could have been used: group 6 (line graph of raw data), group 7 (bar graph of average), and group 8 (bar graph that shows the average and range). This ordering begins with the most widely used representation (a line graph) that uses raw data and ends with a representation (bar graph) that shows both the average height as well as the range of heights at all time points, a sequencing that would help with the goal of accessibility.

Connecting

Finally, the teacher can help students to draw connections between their responses and other students' approaches as well as connections to the key ideas in the lesson. Rather than having discussions consist of separate presentations of different ways to respond to a particular problem, the goal is to have student presentations build on one another to develop powerful ideas.

Let's suppose that in the Case of Kelly Davis the sequencing of student presentations was group 6, group 7, and group 8, as discussed above. Students could be asked to compare the responses of groups 6 and 8 to see how they are the same and how they are different. This move could highlight the fact that if you fit a line to the top and bottom of the bars in group 8's poster, you would get graphs very similar to the tallest and the shortest plant represented in group 6's poster. Students might also notice that while the range can be seen in both graphs, it is easier to determine it from group 8's graph. Students could compare group 8's poster with group 7's poster to see that while both show the average height of the plants, group 8's poster shows the entire range of values without showing every single value (as shown in group 6's poster).

It is important to note that the five practices build on each other. *Monitoring* is less daunting if the teacher has taken the time to *anticipate* ways in which students might approach a task. Although a teacher cannot know with 100 percent certainty how students will engage with a particular task prior to the lesson, many approaches can be anticipated and thus easily recognized during monitoring. A teacher who has already thought about the science concepts and ideas represented by those solutions can turn his or her attention to making sense of those approaches that are unanticipated. *Selecting, sequencing,* and *connecting,* in turn, build on effective *monitoring*. Effective monitoring will yield the substance for a discussion that builds on student thinking, yet moves assuredly toward the goal of the lesson.

Investigating the Five Practices in Action

Above, we presented the five practices for orchestrating a productive discussion and considered what Ms. Davis' class *might* have looked like had she engaged in these practices and how use of the practices in advance of and during the lesson *could* have had an impact on students' opportunities to learn key science ideas. In this section, we analyze the teaching of Nathan Gates, a seventh-grade teacher who has spent several years trying to improve the quality of discussions in his classroom.

The vignette that follows provides an opportunity to consider the extent to which the teacher appears to have engaged in some or all of the five practices before or during the featured lesson and the ways in which his use of the practices may have contributed to students' opportunities to learn.

ACTIVE ENGAGEMENT 2.2
Read the vignette "Growing Wisconsin Fast Plants: The Case of Nathan Gates" and identify places in the lesson where Mr. Gates appears to be engaging in the five practices. Use the line numbers to help you keep track of the places where you think he used each practice.

Growing Wisconsin Fast Plants: The Case of Nathan Gates

In Mr. Gates's seventh-grade life science class, the early units of the course focus on natural variation and patterns of growth in organisms. In order to study these patterns and variation, students were gathering data on the growth of Wisconsin Fast Plant *(Brassica rapa).* At the end of this lesson arc, Mr. Gates wanted his students to understand three scientific ideas:

1.   Natural variation exists in any population of organisms. To identify patterns and correlations, one needs to use mathematical tools that make it possible to describe "typical" growth (including the spread of values that can be considered typical). Typical growth in Fast Plants is described by range and shape. This is often the case in populations of organisms.

2.   Fast Plant growth is characterized by an S-shaped growth curve, where stem length increases slowly for the first ten to twelve days and then increases quite steeply for about seven more days. Following pollination (around day 18), the stem growth slows considerably.

3.   The growth patterns of Fast Plants can be explained by considering where the plant is "spending" its energy resources at various stages of its life cycle and how that is advantageous (e.g., following pollination the plant does not invest energy resources in additional flower production or stem growth, but instead uses its energy to nurture the growth of seed pods and seeds).

In addition to these learning goals, Mr. Gates also developed performance goals to aid in assessing students' progress during and after instruction. His performance goals for this lesson were:

1.   Students will be able to analyze their Fast Plant data in order to identify typical growth in their plants.

2.   Students will be able to draw a graphical representation of their Fast Plant data by identifying the typical growth pattern of their plants.

3.   Students will be able to describe the growth of a typical Fast Plant by identifying a range of stem length and a general shape of the growth curve.

In preparation for this unit and in consideration of time, Mr. Gates planted Fast Plant seeds in containers to allow time for seed germination. He planted six plants in each container. On day 10, the students received individual plant containers. Students decided to measure "growth" of the plants every two to three days for eleven days, marking a piece of string to indicate the plant height and then putting the string on a ruler to get the height in centimeters. Once students had finished collecting data on the plants, Mr. Gates wanted them to create a representation for their data that would enable them to answer the question: "How would we describe the growth of a typical Fast Plant?"

Mr. Gates told his students that they could represent their data any way they wanted. He also told them they could use their raw data (their actual recorded

40   values) or transform their data in some way, which would be depicted in the representation. He emphasized that students needed to be able to explain: (1) what values they plotted; (2) how they got those values; and (3) why their representation helps to answer the question, "How would we describe the growth of a typical Fast Plant?" In this first discussion about the Fast Plant data, he hoped to
45   focus primarily on learning goals 1 and 2.

As students worked on the task in their groups, Mr. Gates circulated among the eight groups, made note of the different approaches the students used and asked clarifying questions. In addition, he pressed students to think about what information they needed to create their representations, why they chose to
50   represent their data the way they did, and how they could describe typical Fast Plant growth using their representation.

Mr. Gates noted that the groups were using different approaches to represent their data—different formats (bar graphs, pictures, line graphs) and measures of central tendency (e.g., mean, range). He thought that group 1 had the
55   most unusual approach of all, choosing to represent their data by creating pots for each plant indicating the length of each plant in the pot at the indicated time points. Mr. Gates noticed that although this approach provided information about plant height, there might be some difficulty in interpreting the representation.

60   After about thirty minutes of small-group work, Mr. Gates decided that it was time to begin a discussion of the students' work. He reviewed his notes that indicated what each group had done:

Group 1	*picture of six pots that show the height of the plant at each time point (plants not drawn to scale)*
Group 2	*line graph that shows the height of four plants at each time point—the two "extreme" plants were not included*
Group 3	*line graph that shows the height of only two plants at each time point*
Group 4	*bar graph that shows the height on the horizontal axis and the number of plants on the vertical axis*
Group 5	*box-and-whiskers plot that shows the range and the median for each time point*
Group 6	*bar graph that shows the shortest and tallest plant at each time point*
Group 7	*line graph that shows the height of each plant at each time point*
Group 8	*bar graph that shows the average height of the six plants at each time point*

65

70

75

Although he instructed each group to hang their poster on the wall, he quickly decided to focus the discussion on the representations produced by

80   group 7, group 1, group 8, and group 5. He felt that this set would highlight
a range of approaches for representing the data and, he hoped, make clear
that some representations provided more insight into typical plant growth
than others.

He began by asking Ryanne from group 7 to share her group's work with
85   the class. Since three of the groups had produced line graphs, this seemed like
a good place to start. Although there were four members of the group, it had
been a few days since Ryanne had shared ideas during a whole-class discussion,
and Mr. Gates wanted this student to have an opportunity to demonstrate her
understanding.

90   Once Ryanne reached the front of the room, she explained that her group
measured the height of each plant and found that from day 13 to day 21 the
plants grew a lot. So, she explained, they chose to represent their data in a line
graph that depicted the growth of all six of their Fast Plants in a different color
as shown in figure 2.3.

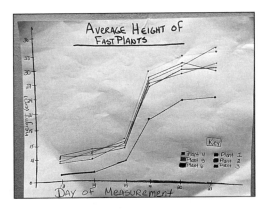

Fig. 2.3. The line graph produced by group 7

95   Mr. Gates then posed a question to the class: "What are some things you notice
about the representation group 7 has created?" Several students shared their
ideas:

Juan: You can easily see the day of measurement and the height of the plants.

Mr. Gates: Okay, Juan, where do you see that?

100   *Juan:*   The graph has axes that are labeled and there is a key so we can tell which
plant is which.

Mr. Gates: Okay, so the *x* and *y* axes allow us to understand what data is represented.
Class, do we agree with that?

Trina: I do, and you can also see the height of all the plants on any day they were
105   measured.

Mr. Gates:	Okay, so what does this graph tell you about the plants' growth?
Trina:	The plants get taller over time.
Mr. Gates:	Okay, the plants get taller over time. What else?
David:	Some plants are growing faster and taller than others.
Tessa:	The plants start out growing slowly, then they really grow a lot, and then they sort of don't grow much.

At this point, Mr. Gates asked the class if they could "see" what Tessa described in the graphs. Marcela, from group 8, volunteered, "Each of the graphs has the same basic shape that sorta looks like an S." Mr. Gates asked the class whether the line graphs that groups 2 and 3 had produced (which were displayed for all to see) had this same general appearance. The students all nodded in agreement. Moses, from group 3, commented, "Yeah, no matter whether it's a tall plant or a short plant, it still has the same shape." Mr. Gates noted, "So, could we say that an S-shaped growth curve is typical for Fast Plants?" Many students again nodded their agreement. England added, "You can really see from all the line graphs that the plants have an S-shaped growth curve over the time that we measured them." Mr. Gates explained that it was typical for these plants to grow slowly at the beginning of their life cycle, followed by a steep increase in growth that can be seen in these graphs. Although the idea *typical growth* had not been specifically raised by the first group, by building on what Tessa had noticed about the plants, Mr. Gates was able to get students to consider an S-shaped growth curve as a way to describe typical growth.

Mr. Gates thanked Ryanne for her contribution, and then asked Peter from group 1 to explain his group's representation (shown in fig. 2.4). Peter explained that their group also chose to represent the height of each plant just like Ryanne's group, but they didn't make a graph. He explained that each pot represented a plant and that each of the stems in the pot represented the height the plant was on a certain day.

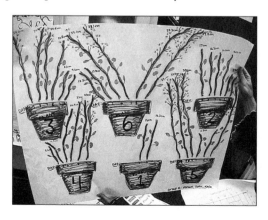

Fig. 2.4. The plant drawings produced by group 1

135   Mr. Gates asked the class what they noticed about group 1's graph. The
following exchange unfolded:

Mitch:	Um, well, it's hard to see the measurements and hard to compare the plants.
Mr. Gates:	Okay, Mitch says that it is hard to see the measurements and difficult to compare the plants with this representation. Does anyone else agree?
140	*Students:*	Yes!
Mr. Gates:	Okay, Marie, you said yes, what makes this representation difficult for you to understand?
Marie:	I think the pots and the leaves and the plants make it really confusing. The days aren't in order and I don't think the stems in the pots are equal to the
145		real pots. So it is hard to compare across pots. It is really unorganized.
Mitch:	I agree. Ryanne's group's graph is easier to read. We can see stuff easily. You can see how tall each plant is each day they measured and you can compare the heights of the plants on the same day.
Mr. Gates:	Okay, Mitch says we can see stuff easily in the graph. What does that tell us
150		about representing scientific data?
Peter:	I guess graphing our data would have made it easier for everyone else to see. And, they would have an easier time making comparisons from one day to the next. But it's pretty the way we did it! And you can see other stuff like when we got flowers.
155 | *Mr. Gates:* | Okay, Peter, great points. You were showing more information than just height because you drew the flowers in, too. And, it might be very easy for you to see and understand your own data in a representation like this, but it can be difficult for others to interpret. This very idea is why in science it is important to use standard representations, like a line graph or even a bar |
160 | | graph, to represent data. It allows us to easily see and interpret the data, especially when it is data that we didn't collect. |

Mr. Gates then commented that several groups, including groups 1 and 7,
graphed all their data. He indicated that graphing all the data is a great way to
represent all the information but that if there were a lot of data it might become
165   confusing. He then explained that some groups chose to represent their data a
little differently. He called on Tristan to explain the representation produced by
group 8 (shown in fig. 2.5).

Tristan explained that they thought it would be easier to take the average
of all six plants in order to get the "typical" growth. So, they figured the mean
170   plant height for each day and created a bar graph to show the means. Tristan
said that for day 10 the average plant height was 10.32 cm and the mean
increases from there.

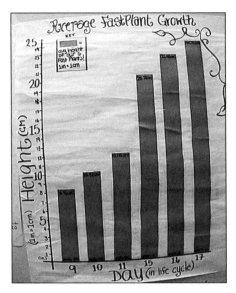

Fig. 2.5. The bar graph from group 8

Mr. Gates thanked Tristan for his explanation. He then asked for a volunteer to explain what the mean is and how group 8 found it. NeeNee explained that you find the mean by adding the heights for all the plants and dividing that number by the total number of plants. Mr. Gates then asked if someone could explain what the mean tells us. Allison volunteered, "The mean tells you the average height of the plants. Like, on that day the plants can be expected to be around that number. You use the actual data and calculate the mean." The discussion then continued:

Mr. Gates:	Thank you, Allison. Did everyone hear her? She said that mean is a measure that is calculated based on the raw data. Now, what do you think might happen to group 8's data if on day 10, one plant started to grow really fast and was much taller than any other plant?
Phaedra:	So, what happens if there is a really tall plant in the pots compared to all the others?
Mr. Gates:	Yes, what might happen to the mean if there happened to be a really tall plant compared to all the other plants? Would the mean be any different?
Phaedra:	Well, with even just one really tall height, you would have a bigger sum when you add all the heights together, so the mean would be bigger, too.
Mr. Gates:	Let's think about what Phaedra just said. She said if one of the plants were much taller than the rest, the mean would increase. How do you feel about this idea, Mikhail?
Mikhail:	Well, it makes sense that having a larger number increases the mean. If one

195   of the numbers changed from 9 to 29, that's a 20-cm difference. That's huge. The mean would definitely be bigger.

Karen: Yeah, but if you had a really tall plant, then it might not look like the rest of them. The mean might not tell you what a typical plant looks like.

Tristan: You mean using the mean might not tell you what a typical plant looks like?

200   *Karen:* Yeah.

Mr. Gates: Okay, so if we had a really tall plant, or even a plant that is really short, that would influence the mean. When we have data that are really different from the other data we call them outliers. Outliers can distort the mean.

Patrice: So are you saying that the mean is not a good thing to use if we want to

205   describe a typical plant?

Mr. Gates indicated that Patrice has asked a really important question— how do you describe a *typical* plant? He explained that the mean provides valuable information but that you just have to be aware of the outliers. He asked Katie from group 2 to explain what they did. Katie said, "We were wor-

210   ried about the fact that we had one plant that was really tall and one that was really short and the other four were very close together. So we just graphed the four that were close together and thought that any one of them could be considered typical."

Mr. Gates said that group 5 used an approach that used all the data but

215   tried to deal directly with the issue of typicality. He called on Bri from group 5 to explain their representation (shown in fig. 2.6). Bri explained that her group thought it might be important to show the variation in height among the plants for each day measured and to show where the median height was on each day, and so they decided to make a box-and-whiskers plot. Mr. Gates asked Bri to

220   explain the plot and her group's thinking behind it. Bri explained that if they wanted to know what is typical for a Fast Plant on day 14, for example, she could tell them that the heights ranged from 12 to 26 cm, that the median was 19 cm, and that 50 percent of the plants had heights between 16 and 24 cm.

Mr. Gates commented that showing a range of data could be very helpful

225   in describing typical growth. He explained that because every organism, every plant, is different, heights and growth vary, but there is an expected height that we can see for each day. In other words, he explained, we can expect that most Fast Plants would fall within a certain range of heights in their growth cycle.

Mr. Gates asked the students what Bri and her group needed to do in order

230   to convert their data into a box-and-whiskers plot, because he wanted to make sure they understood both how to create the plot and how to read it. Students discussed ordering the data for each time point, finding the low and high values (the whiskers), and determining the median, as well as the values that separated the top quarter and bottom quarter (edges of the box).

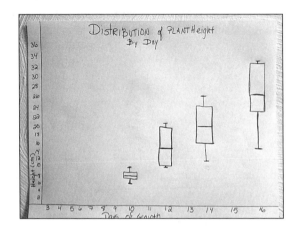

Fig. 2.6. The box-and-whiskers diagram from group 5

235   With only a few minutes left in the class, Mr. Gates told students that for
homework he wanted them to answer the question "How can you account for
the S-shaped growth curve?" In other words, why might these plants typically
grow in this way? He told students that he expected a written answer to the
question and a rationale for their conclusions. He decided to use this question
240   to launch a discussion in the next class, and hopefully make progress on learning
goals 2 and 3.

Analyzing the Case of Nathan Gates

Although we could identify many aspects of the instruction in Mr. Gates's classroom that may have
contributed to his students' opportunities to learn, we will focus our attention specifically on his use
of the five practices. In subsequent chapters, we will analyze a broader set of actions that, in combina-
tion with the five practices, help account for the success of the lesson. We will begin by considering
the five practices and whether there is evidence that Mr. Gates engaged in some or all of them. Then
we will consider how his use of the practices may have enhanced his students' opportunities to learn.

Evidence of the Five Practices

As we indicated in chapter 1, determining clear and specific goals for the lesson and selecting a
task that aligns with the goals are the foundation on which the five practices are built. Hence, Mr.
Gates's identification of the three scientific ideas that he wanted his students to learn (lines 6–19)
and his selection of a task that had the potential to reach these goals (lines 31–37) positioned him
to use the Five Practices model effectively.

Anticipating

Because the vignette focuses primarily on what happened *during* a classroom episode, we have limited insight into the planning Mr. Gates engaged in prior to the lesson and the extent to which he
anticipated specific solutions to the task. However, the fact that he wanted students to know that

the typical Fast Plant growth is described by shape and range suggests that he had considered the possibilities for representing the data that would highlight both of these attributes—namely, line graphs and box-and-whiskers plots, respectively. In addition, Mr. Gates's decision to begin the next class with a discussion of student responses to the question "How do you account for the S-shaped curve?" (lines 235–37) suggests that he considered how particular representations (such as line graphs) would help in accomplishing this goal for the lesson.

Monitoring

Mr. Gates monitored students working in their small groups (lines 46–59). Through this monitoring he was able to determine the approaches that specific groups were using (lines 46–48; 52–57), ask questions to help students make progress on the task (lines 48–51), and recognize that the representation used by group 1 was difficult to interpret (lines 57–59). His monitoring of the students' work provided information about their thinking that he needed in order to make decisions about which representations to focus on during the discussion.

Selecting

By referring to notes he had made during the monitoring process (lines 61–77), Mr. Gates knew which groups had produced specific representations. Armed with this information, he decided to have particular groups (7, 1, 8, and 5) present posters that would highlight different information about the plants, thus providing grist for the discussion of what needs to be considered in determining what is typical (lines 78–83). In addition, he decided that he wanted students to consider the representation produced by group 1 (see fig. 2.4) so that they could see that some representations made it challenging to identify patterns and correlations in the data.

Sequencing

Mr. Gates selected Ryanne from group 7 as the first presenter, since the representation produced by her group (line graph) had been used by several groups and therefore was likely to be one to which other students in the class could readily relate (lines 84–86). In addition, he wanted to give Ryanne a chance to participate actively and publicly in class (lines 86–89) as it had been several days since she had done so. By selecting Ryanne, Mr. Gates was able to both highlight a popular strategy and make sure he was providing his students with equitable opportunities to demonstrate competence.

While Mr. Gates doesn't explain precisely why he chose to have groups 1, 8, and 5 present their posters in that order, we might infer his intent from the way the discussion unfolded. Specifically, he started with a graph (fig. 2.3) that was similar to ones produced by other students in the class. This graph portrayed all the data that had been collected, allowed for comparisons across plants, and showed the shape of the curves (an important feature in explaining how the plants use their resources over their life cycle). He next selected a representation that also used all the data but that made any type of comparison challenging. This highlighted for students the need to use standard representations (lines 156–161).

Mr. Gates then asked group 8 to present. The graph produced by this group (fig. 2.5) was different from the others in that it featured only the average height of the six plants at each measurement point rather than all of the data. This provided an opportunity to talk about the mean, how outliers might affect the mean, and how the mean might not be the best measure of what is typical (lines 173–207). Mr. Gates concluded the presentations with a discussion of the poster created by group 5 (fig. 2.6). This poster had some of the features of those presented earlier—it used a measure of central tendency (like group 8), and it showed the range of values (like group 7)—but also some important differences. First, the group used the median as a measure of "average" instead of the mean. Since outliers do not affect the median, this value separates the data into the top and bottom 50 percent. In addition, the low and high values are included so the range of data can be easily determined. The representation also made clear where most of the data fell—between the top and bottom edges of the box. Hence, Mr. Gates was able to make the point that showing a range of data can be helpful in describing typical growth (lines 224–28). He may have decided to end the discussion with this poster because it brought together several ideas that had been discussed in earlier posters and that were important to understanding the growth of Fast Plants, and it was a more sophisticated strategy that might not have been accessible to all groups without first analyzing simpler graphs.

Connecting

Through the questions that Mr. Gates asked during the discussion and the ways in which he pressed students to clarify what they had done and why, he helped students make connections with the scientific ideas that were the target of his instruction. Specifically, Mr. Gates indicated that he wanted his students to understand three scientific ideas: (1) Typical growth in Fast Plants is described by range and shape; (2) Fast Plant growth is characterized by an S-shaped growth curve, where stem length increases slowly for the first ten to twelve days and then increases quite steeply for about seven more days; and (3) The growth patterns of Fast Plants can be explained by considering where the plant is "spending" its energy resources at various stages of its life cycle and how that is advantageous.

Mr. Gates pressed students to "see" if they could describe the phenomena that was articulated by Tessa (lines 112–13), which resulted in the identification of the basic S shape and the realization that the same persisted regardless of the height of the plant. In addition, through the analysis of several graphs, Mr. Gates was able to highlight the point that Bri made in describing group 5's poster: Showing a range of data is important (lines 224–25). By questioning students about the carefully sequenced work he was able to help them understand two key ideas (learning goals 1 and 2) that were targets for the lesson. While no progress was made on goal 3 during the course of this one lesson, the assigned homework was intended to serve as a launching point of this conversation the following day.

Mr. Gates did not make explicit connections among the student graphs in the lesson. However, during the discussion of group 1's poster, Mitch spontaneously referred to group 7's graph, commenting that it was easier to read (lines 146–48), and Peter added more specificity to the discussion by indication that group 7's poster made it easier to make comparisons (lines 151–54).

Relating the Practices to Learning Opportunities

Did Mr. Gates's use of the five practices contribute to his students' learning? Although we have no direct evidence of what individuals in the class learned, we see a group of students who appear to be engaged in the learning process. Over the course of the lesson, the teacher involved fifteen different students (half of the students in the class) in substantive ways. Mr. Gates repeatedly targeted key ideas related to the goals of the lessons as he guided his class in discussing four different representations in some depth. The final question that he gave for homework (lines 235–37) provided individual students with an opportunity to make sense of what had transpired during class and to make connections that would provide the teacher with insight into their thinking.

The Five Practices model gave Mr. Gates a systematic approach to thinking through what his students might do with the task and how he could use their thinking to accomplish the goals that he had set. Although we analyzed the practices in action—what the teacher did during the lesson—we argue that to do what he did during the lesson, he must have thought it all through *before* the lesson began. We will explore how to engage in such planning in subsequent chapters.

The Science Practices in the Case of Nathan Gates

On page 1 in the introduction (fig. 0.1), we listed the eight Science Practices set forth in the Next Generation Science Standards (NGSS) (Achieve, Inc. 2013). Here, we analyze the opportunities Mr. Gates's students had to productively engage in many of these science practices. As the lesson began, students had an opportunity to plan and carry out an investigation (SP 3) as they measured the growth of their Fast Plants (lines 32–34). During their investigations, students collected and represented data on plant height (lines 34–41). As students constructed their representations, they interpreted and analyzed their data (SP 4) in various ways, including using measures of central tendency (lines 52–54). Additionally, Mr. Gates asked the students to construct an explanation (SP 6) detailing how their representation answered the question that guided the investigation (lines (48–51). In designing the task in this way, Mr. Gates provided students with an opportunity to use mathematics and engage in computational thinking (SP 5) as they described the growth of a typical Fast Plant.

Once the whole-class discussion began, we saw students asking questions (SP 1) about typicality and how outliers affect the mean stem length (lines 185–86, 199, 204–5). Furthermore, as each student communicated his/her group's findings (lines 90–96, 130–34, 168–73, 216–32), Mr. Gates prompted students to critically examine and evaluate the work of their classmates (SP 8). Finally, the homework Mr. Gates assigned gave students an opportunity to use what they learned during the task and whole-class discussion to construct an explanation (SP 6) about the Fast Plants' growth based on the S-shaped curve (lines 235–37).

The way in which Mr. Gates selected, designed, and implemented the Fast Plants task provided multiple opportunities for students to engage in the NGSS Science Practices. In doing so, students talked productively with one another about the science content, which focuses on the NGSS disciplinary core ideas MS-LS1.B (Growth and Development of Organisms) and MS-LS1.C (Organization for Matter and Energy Flow in Organisms) (Achieve, Inc. 2013) through a detailed analysis of their data and variety of representations.

Conclusion

Mr. Gates avoided a show-and-tell session in which solutions are presented in succession without much rhyme or reason, often obscuring the point of the lesson. By carefully considering the story line of his lesson—what he wanted to accomplish and how different representations would help him get there—he was able to skillfully question his students and position them to make key points. So, with the lesson always firmly under his control, the teacher was able to build on the work produced by students, carefully guiding them in a sound direction.

The Case of Kelly Davis discussed in the beginning of this chapter provides a contrast to Mr. Gates's instructional approach. Although the students in Ms. Davis's class used a range of interesting approaches, what the students were supposed to learn from the sequence of presentations was not clear, other than that "the data can be represented in many different ways." The students took with them no clear understanding about science concepts and ideas from this experience.

The five practices build on each other, working in concert to support the orchestration of a productive discussion. It is the information gained from engaging in one practice that positions the teacher to engage in the subsequent practice. For example, a teacher cannot select responses to share in the whole-class discussion if she or he is not aware of what students have produced (the teacher needs to monitor to be able to select and sequence). And a teacher can't make connections across strategies and to the goal of the lesson if she or he has not first selected and sequenced strategies in a way that will help advance the storyline of the lesson. In the next two chapters we explore the five practices in more depth, building on the descriptions provided in this chapter.

Getting Started: Anticipating and Monitoring Students' Work

O nce teachers have identified a learning goal and selected an appropriate task aligned with that goal (as discussed in chapter 1), they are ready to begin work on the five practices. In this chapter, we discuss the first two practices—*anticipating* and *monitoring*—and consider what teachers can do prior to and during a lesson to position themselves to make productive use of the student work (artifacts) produced during collaborative activities. By closely examining how one teacher engages in the first two practices, we will show how use of these practices prior to and during a lesson can set the stage for a productive discussion in the concluding phase of the lesson. We will first describe each practice and then present part of the case of a middle school teacher, Kendra Nichols, as she uses that practice. We will conclude with an analysis of how Ms. Nichols's use of each practice leads to productive discussion in her classroom.

Anticipating

Anticipating involves carefully considering: (1) the key features that must be present for a complete and correct experimental design, explanation, or representation; (2) the challenges that students are likely to encounter and/or the misconceptions that might surface as they engage in the task; and (3) how to respond to the work that students are likely to produce that may or may not address the identified features. We will use the vignette Matter and Molecules: The Case of Kendra Nichols (Part 1—Anticipating) to illustrate these three points.

ACTIVE ENGAGEMENT 3.1

- Imagine that you have engaged middle school students in a series of exploratory activities where they noted patterns in the behavior of water (as summarized in fig. 3.1). Following these explorations, you asked the students to explain those patterns by imagining how the particles of water might look and behave (fig. 3.2). What are the key features you would be looking for in a complete and correct explanation of the patterns noted in the behavior of water?
- What conceptions or challenges do you think are likely to surface in students' representations and explanations?

Matter and Molecules: The Case of Kendra Nichols
(Part 1—Anticipating)

Ms. Nichols recently started a unit on Kinetic Molecular Theory (KMT) in her eighth-grade integrated science class. Over the course of the unit, which would take several weeks to complete, she wanted her students to understand a set of disciplinary ideas that were reflected in her learning goals:

5   1. All molecules are constantly in motion.

2. States of matter are characterized by different molecular motion:

   a. *Solid:* molecules vibrate

   b. *Liquid:* molecules move randomly within limits

   c. *Gas:* molecules move randomly with no limits

10   3. Transitions between solid, liquid, and gaseous phases typically involve large amounts of heat energy. In the case of water, to transform solid water (ice) into liquid water, you need to add heat energy (which causes the ice to melt). To transform liquid water into gaseous water, you need to add heat energy (which causes the liquid water to boil). The opposite is also true:

15   condensing gas to liquid and liquid to solid (freezing) requires a loss of heat.

4. When you add heat energy to a substance, the molecules of the substance move more and faster.

5. Increased molecular motion moves molecules farther apart (in almost all

20   substances).

6. Water is an unusual substance because the molecules of the solid are farther apart than the molecules of the liquid. This happens because the hydrogen bonds in water are most stable in a rigid array that includes space between the molecules (to minimize the forces resulting from slightly like-charged

25   particles repelling one another).

Ms. Nichols decided to begin the unit by having students experiment with and then explain the behavior of water. Although water behaves differently than other matter, she selected it as the starting point because it is ubiquitous, students have experience with water in different states (solid, liquid, and gas), and

30   it is a substance that be easily and safely explored in all three states in the classroom. In the first task, students engaged in four experiments with solid, liquid, and gaseous water and identified patterns of how water molecules interact at varying temperatures. Ms. Nichols posted the patterns (fig. 3.1) in the classroom so students could refer to their findings from these experiments during subse-

35   quent work on the Explaining the Behavior of Water task (an "explanation" task, as described in chapter 1), shown in figure 3.2. Ms. Nichols decided to use this task next because it provided a context for explaining the molecular makeup of

water and for making sense of the unique nature of water molecules. In addition
to her previously established learning goals, Ms. Nichols determined that the
40 performance goal for the lesson would be for students to represent the molecular
configuration of water in each phase and that the representations would include
the depiction of motion, spacing, and force and heat energy. Through this lesson
she hoped to address learning goals 1, 2, 3, 4, and 5. All of the learning goals,
including learning goal 6, would be revisited as additional substances were ex-
45 plored in subsequent lessons.

Experiment	Description	Patterns They Noticed
1	Observation of physical properties of solid, liquid, and gaseous water.	Solid water retains a definite shape and can't be compressed using a syringe. Liquid water flows to the bottom of the container and is not measurably compressed in the syringe. Gaseous water escapes (disperses) out of the container and can be compressed when put into a syringe (when we push down on the plunger).
2	Measurement of mass and volume of vials of water at 20°C (time 1) and then again once the temperature was lowered or raised. . . . At time 2, the vials were at 0°C (freezing), 2.8°C (refrigerator temperature), or 20°C (room temperature).	The mass of the water in our 15 vials didn't change from time 1 to time 2, but the volume of water in the vials we placed at 0°C increased about 10%. The volume of the water in the vials at 2.8° C and 20°C didn't change measurably.
3	Observation of behavior of ice in liquid water.	Ice floats on liquid water, but not all the way at the top. Part of the ice cubes are under the surface of the water and part of them are above the surface.
4	Observation of boiling and condensation.	When heat is added to the flask, the water level decreases. Small droplets of water collect in the tubing (and we can also see "fog" in the tube) and then droplets form in the test tube that is sitting in the cup of ice.

Fig. 3.1. Patterns students identified in their four experiments with water

Imagine that you have a pair of Magic Science Glasses. You put on your Magic Science Glasses and you look at the glass of water. But instead of the water, you see the pieces that make up the water, the particles.

I want you to put on your Magic Science Glasses and "look" at the solid, liquid, and gaseous water that we've been experimenting with. I want you to use what you see about how these particles are behaving and what they look like to explain all those patterns that we noticed in our experiments.

Keep our patterns table in mind as you do this task. In your groups, you are going to DRAW a representation of what you see when you put your Magic Science Glasses on. Use the whiteboards to do this. I'll give you twenty minutes and then we are going to share our ideas and see if we can come up with one explanation to account for what we saw.

Fig. 3.2. Explaining the Behavior of Water task

Ms. Nichols began planning the Explaining the Behavior of Water lesson by reviewing the information provided in the Matter and Molecules curriculum materials (Institute for Research on Teaching 1998) and referring to the NSDL Science Literacy Maps (found at http://nsdl.org) in order to familiarize herself with possible approaches her students might use in drawing and explaining their representations. For this particular lesson, she developed measureable performance goals, which she would use to assess students' throughout the activity. The performance goals for this lesson were as follows:

1. Students will be able to depict the molecular behavior of water in all three phases in drawings on whiteboards.

2. Students will be able to explain the molecular behavior of water in the solid, liquid, and gaseous phases.

Based on her reading, and her experiences with her students, she identified key features that needed to be included in the explanations and/or representations:

- Water molecules are further apart in the solid than liquid phases (spacing);

- Molecules are moving in all phases (motion);

- Heat energy is needed to transform molecules from solid $\rightarrow$ liquid $\rightarrow$ gas; heat energy must be removed to transform molecules from gas $\rightarrow$ liquid $\rightarrow$ solid; and

- Solid molecules are held together by forces in an organized array that minimizes repulsion between atoms.

Ms. Nichols thought about the general understanding of molecules that most of her students had at this level (e.g., that liquid and gas molecules move but solid molecules do not and that solid molecules are more tightly packed than liquid or gas molecules) and how they would need to move beyond these understandings in order to successfully complete the task. Hence she expected that students would be confused about or fail to attend to some of the key features she had identified, and she wanted to make sure that she was prepared to support them in developing coherent explanations based on patterns noticed in the earlier lessons. For example, she thought that many students would have problems with molecule spacing. Because of water's unique nature (the molecules of the solid take up more space than the same amount of molecules of the liquid), she wanted to be sure to stress the idea that molecules in solid water are spaced further apart than molecules in liquid water. She planned to have jars of marbles available, and when the need arose, she would ask students if they would be able to put more or fewer marbles in a jar if they could push the marbles closer together and pack them tighter. Once the students realized that they would be able to put more marbles in the jar, she would ask them to explain what happened during experiment 2 (see fig. 3.1; volume of water increased when frozen) to get students to understand the concept of molecule spacing. She also decided that having some magnetic marbles available would help students think about forces and how this would impact the spacing of the water molecules.

Ms. Nichols thought that many students would have trouble identifying the role of heat—it must be added or removed from the system in order for the water molecules to change states. She made note that she would ask the students to explain what happened in experiment 4 (see fig. 3.1). In this experiment, students noticed that when heat was applied to the flask, the level of water decreased, but water droplets collected in the tubing away from the heat source. Getting students to explain condensation and evaporation of the water would help them to understand that increasing heat energy results in changes of phase (e.g., from liquid to gas) and decreasing heat energy results in changes of phase (e.g., from gas to liquid). Revisiting experiment 4 would also raise the issue of motion—the water started out in one flask and ended up in the test tube on the other side of the apparatus. This should help students to realize that the molecules had traveled through the tubing.

In addition, the Matter and Molecules curriculum identified two additional ideas that students might bring to the task: (1) that molecules in different phases were different sizes; and (2) that solid molecules are rigid and liquid molecules are flexible—that is, features of the macro substance are assigned to the molecules themselves. Ms. Nichols decided these were also things that she needed to be on the lookout for.

Finally, she anticipated that some students might struggle to even begin the task. Instead of telling these students exactly what they needed to do, or asking specific prompts to lead them to develop the desired representation, she wanted to give students the opportunity to develop their own representations based on the patterns they noticed in the initial experiments (see fig. 3.1). She decided she would ask students a series of questions that would provide them with the opportunity to summarize the patterns in their own words and then prompt

115   them to create a drawing that shows what water molecules look like based on those patterns.

Ms. Nichols wanted to make sure that when she got to the end of the lesson, she had accomplished what she set out to do (see the learning goals in lines 5–25). She created a complete and correct representation that incorpo-

120   rated the features she had identified as a reference for herself (fig. 3.3). During the whole-class discussion she planned to either use a representation produced by students or have the whole class jointly construct an explanation in order to make salient the disciplinary ideas she was targeting in the lesson.

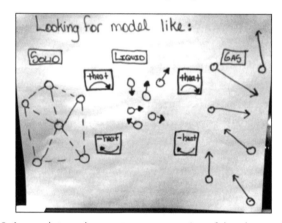

Fig. 3.3. A complete and accurate representation of the phases of water

Because she wanted as much as possible to build the whole-class discus-

125   sion on the work students produced, Ms. Nichols decided to keep track of what students were doing as she observed and interacted with them while they worked on the task in small groups. To facilitate this, she made a chart (fig. 3.4) where she could record key features of the explanations produced by the groups, including what features were missing or incorrectly portrayed in their repre-

130   sentation. She thought the chart would help her in planning the whole-class discussion of the representations and related explanations, specifically in think-ing about which explanations to feature and the order in which they should be discussed. In addition, she thought the chart would help her keep track of "where students were" in their understanding of the central ideas of the lesson.

STUDENT GROUP	Features Correctly Represented	Features Missing or Incorrectly Represented	ORDER and NOTES
	__ SPACING __ HEAT __ MOVEMENT __ FORCES __ TYPE & SIZE	__ Incorrect Spacing __ No/Incorrect Heat __ No/Incorrect Movement __ Different-Size Molecules __ Other __ Accurate Representation	
	__ SPACING __ HEAT __ MOVEMENT __ FORCES __ TYPE & SIZE	__ Incorrect Spacing __ No/Incorrect Heat __ No/Incorrect Movement __ Different-Size Molecules __ Other __ Accurate Representation	
	__ SPACING __ HEAT __ MOVEMENT __ FORCES __ TYPE & SIZE	__ Incorrect Spacing __ No/Incorrect Heat __ No/Incorrect Movement __ Different-Size Molecules __ Other __ Accurate Representation	

Fig. 3.4. A chart for monitoring students' work on the Explaining the Behavior of Water task

ACTIVE ENGAGEMENT 3.2
Compare your response to Active Engagement 3.1 with the features and challenges Kendra Nichols anticipated.

Analysis of Anticipating in the Case of Kendra Nichols

In the Case of Kendra Nichols, we see a teacher who engaged in thorough and thoughtful planning for a lesson she was going to teach in her eighth-grade integrated science class. Her planning started with the identification of the disciplinary ideas that she wanted her students to learn in the unit (lines 5–25) and the selection of a series of tasks (figs. 3.1 and 3.2) that would provide students with a starting point for learning about molecules and matter. She was clear about what she hoped would be accomplished in the Explaining the Behavior of Water task (lines 38–43) and in subsequent lessons (lines 43–45). Once she determined what students were going to do and why, she turned her attention to anticipating what was likely to happen as students went to work on the task and what she would need to do to support their efforts. We now consider the aspects of anticipating discussed at the beginning of the chapter and consider Ms. Nichols's engagement in each component of this practice.

Key Features That Must Be Present

The task that Ms. Nichols selected for her students (fig. 3.2) required them to create a representation that would explain what they saw when they put on their "Magic Science Glasses." To be positioned to determine the correctness and completeness of the representations and explanations generated by students, Ms. Nichols had to first identify the components or features of a high-quality explanation that could serve as a standard for evaluating student responses (lines 58–67) and create a representation that embodied these features to serve as a reference point (fig. 3.3). She then created a monitoring sheet (fig. 3.4) that highlighted these features and made it possible for her to track the extent to which various features were present in the work produced by each group. The chart could be used for a variety of purposes. By providing a record of who is doing what, the chart can serve as a data source for making judgments about who will share what during the discussion, help the teacher keep track of how students in the class are thinking about particular ideas, and provide a record of which students were selected to share their work on a particular day. In addition, it can provide a historical record of what happened during the lesson that can aid the teacher in refining the lesson the next time it is taught.

Challenges and Misconceptions

Ms. Nichols gave careful consideration to both what her students might know and what they might do. First, she recognized that students brought to the lesson some general understandings about molecules that may or may not be correct (lines 68–72). By being aware of students' prior knowledge, the teacher is better positioned to understand initial steps that students might take and to help them move in a more productive direction.

Second, Ms. Nichols identified a misconception that students might have related to the spacing of molecules (lines 76–80)—that the solid was more densely packed than the liquid or gas. Because this misconception is related to one of the key features of the explanation she was looking for (line 61) it would be a critical idea to identify and address. Finally, Ms. Nichols recognized that some students might have trouble beginning the task without more specific directions or assistance (lines 108–9). Because she recognized that this was a possibility, she was able to develop a plan for supporting struggling students without taking over the thinking for them (i.e., while protecting the cognitive demand of the task; see chapter 1).

Responding to Students' Work

The final aspect of anticipating involves determining how to respond to the work that students produce that may or may not be accurate or complete. Having first identified the key features of a high-quality explanation and the challenges and misconceptions likely to surface as students work, the teacher must now consider what to do to move students toward the disciplinary ideas she has targeted for the lesson while still allowing them to engage in a cognitively challenging science practice.

Ms. Nichols anticipated various strategies for responding to students' work during the Behavior of Water task. First, she indicated that she would provide students with material resources that they could use to make sense of specific phenomena. For example, she planned to have jars of marbles available to help students think about molecule spacing (lines 80–86) and magnetic marbles available to help students think about bonds and how they impact the spacing of water molecules (lines 86–88).

Second, Ms. Nichols planned to have students consider how their prior knowledge (in this case, that gained from the experiments in the first task they engaged in) could be brought to bear in making sense of the current task. For example, she planned to ask students to look at the patterns they identified in experiment 4 in order to identify the role heat plays in water transitioning from one state to another (lines 92–98) and to raise the point that molecules are moving (lines 98–101). Third, Ms. Nichols recognized that some students might have trouble getting started and planned to ask them a series of questions (based on the experiments they had previously conducted) that would help them begin to engage with the task (lines 108–16).

Finally, Ms. Nichols identified specific questions that she would ask during the lesson (lines 80–83; 91–92; 112–16). Although the set of questions that she might ask her students as they worked on the task was somewhat limited (and, in fact, it might have been helpful for her to have identified a few more), having some specific questions ready in advance of the lesson (in her "back pocket," so to speak) meant that Ms. Nichols would not need to develop all the questions on the spot. This would give her more time to consider the appropriate moment to ask a particular question to help to make connections between what students were actually doing and the disciplinary ideas that she wanted them to learn. As one of the teachers with whom we have worked commented:

> Coming up with good questions before the lesson helps me keep a high-level task at a high level, instead of pushing kids toward a particular solution path and giving them an opportunity to practice procedures. When kids call me over and say they don't know how to do something (which they often do), it helps if I have a ready-made response that gives them structure to keep working on the problem without doing it for them. (Smith, Bill, and Hughes 2008)

Because questions are bound to the context in which they are asked, it is critical to pose a question that makes a connection with issues that are currently being addressed. Developing questions only "in the moment" is very challenging for a teacher who is juggling the needs of a classroom full of learners who need different types and levels of assistance. When teachers feel overwhelmed by the needs and frustrations of their students, it is easy for them to revert to just telling students what to do when an alternative course of action does not immediately come to mind.

Monitoring

Monitoring is the process of paying attention to the thinking of students during the actual lesson as they work individually or collectively on a particular task. This is not just listening in on what students are saying and observing what they are doing. It also requires keeping track of the features of the experimental designs, explanations, or representations that are present and absent from their work; asking questions that will help students make progress on the task; and identifying aspects of their work that can help advance the discussion later in the lesson. Such questions might include those that will get students back on track if they are following an unproductive or inaccurate pathway, or that will press students who are on the right course to think more deeply about connections or patterns that they are seeing or about potential underlying causes for these patterns. The teacher's role in the monitoring phase of the lesson is *not* to ensure that each individual or group has a complete and correct response to the task but rather to understand where students are and to push their thinking forward. The whole-class discussion at the end of the lesson is the place where the teacher makes sure that a consensus is built regarding the key features needed for correct and complete experimental design, explanation, or representation and makes certain that the class as a group has an opportunity to learn and discuss the disciplinary core ideas that were driving the lesson. We have found that many teachers want to go further and coach each individual group to mastery during collaborative group work. It is challenging for them to trust that the whole-group discussion can provide an effective context for students to confront their own sometimes incomplete or incorrect ideas and develop that mastery. We address this issue in more depth in chapter 7.

We now return to the Case of Kendra Nichols as she monitors her students' work in an effort to support their learning and engagement with the task and to prepare for the end-of-class discussion.

ACTIVE ENGAGEMENT 3.3

The second part of the Case of Kendra Nichols focuses on the practice of monitoring. As you read part 2—
- identify specific things that Ms. Nichols does to support her students' learning; and
- consider how the data that Ms. Nichols collected in her chart (see fig. 3.6) could be useful to her as she helps students and prepares for the end-of-class discussion.

Matter and Molecules: The Case of Kendra Nichols
(Part 2—Monitoring)

135   Ms. Nichols introduced students to the task (see fig. 3.2) and made sure that they understood what they needed to do. She explained that each group needed to create a poster that included a drawing that accounted for the patterns they noticed in their work on the four experiments (see fig. 3.1) and to also be prepared to explain the ideas. She provided each group with a basket of resources
140   that included markers, a small jar of marbles, and magnetic marbles.

Armed with her monitoring tool—the chart shown in figure 3.4—Ms. Nichols circulated among the small groups and listened in on their conversations. She asked questions as needed to get students on the right track or to press them to make sense of what they were doing as she kept track of who
145   was doing what. For example, when Ms. Nichols approached group B (Kim, Kevin, Lamar, and Rosa), she noticed that the students represented heat in their representation (see fig. 3.5). She asked the group to explain why they did this. Rosa said, "We showed like the melting and freezing and stuff. So it's melting when it goes from solid to liquid. And it's freezing when it goes
150   from liquid to solid. And when it freezes you take away heat. But when you add heat, like when it goes from liquid to gas, you add heat and that is when it boils." While Ms. Nichols was pleased that the group had identified heat as critical in the process, and had accurately described how it impacted the changes between phases, she noticed that they had the misconception that
155   the molecules expanded in size but that the spacing between molecules did not change. She asked Kevin to explain their thinking behind the size of the

molecules. Kevin said, "We showed how the water molecules take up more space when they get hot. So when we did our experiment we said that the gas expanded. And that's what we showed. That when you add heat, it expands." Ms. Nichols asked what they had learned from experiment 2. Lamar said that the vial with frozen water had a bigger volume than the ones that were just sitting in the room or in the refrigerator. Ms. Nichols asked the group to consider if their picture was consistent with this finding and if not, how they might adjust it. She left the group to ponder the question.

<table>
<tr><th>Group</th><th>Representation Drawn</th><th>Group</th><th>Representation Drawn</th></tr>
<tr><td>A</td><td>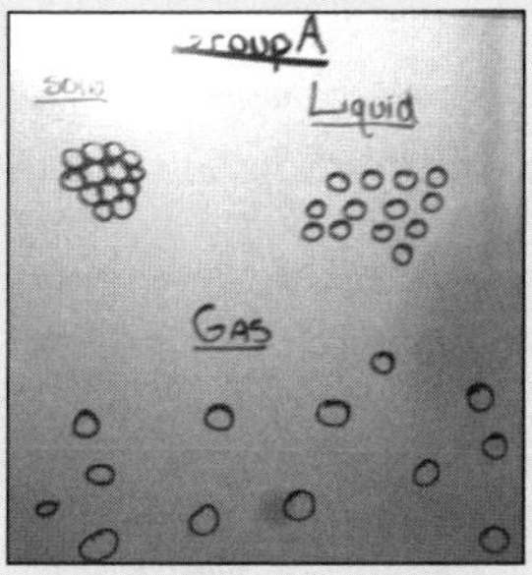
</td><td>D</td><td>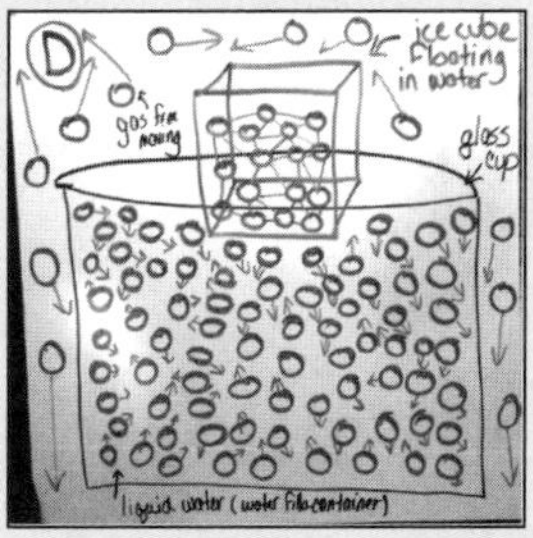
</td></tr>
<tr><td>B</td><td>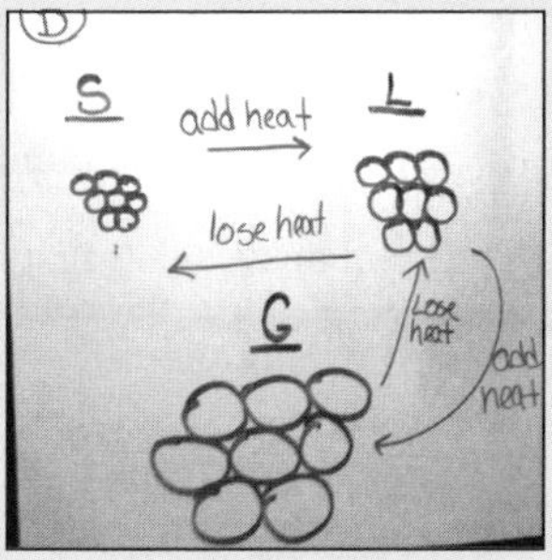
</td><td>E</td><td>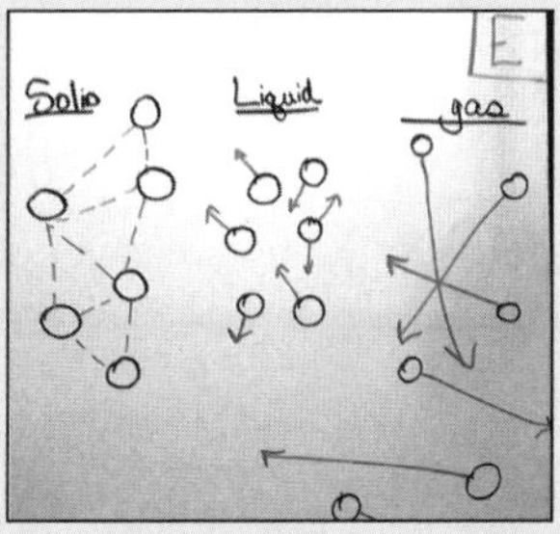
</td></tr>
<tr><td>C</td><td>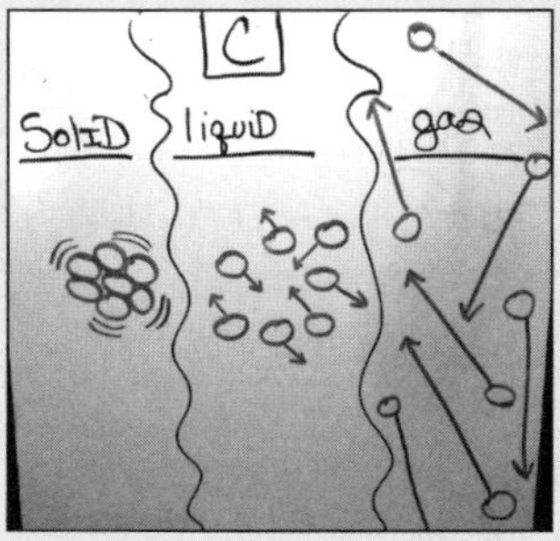
</td><td>F</td><td>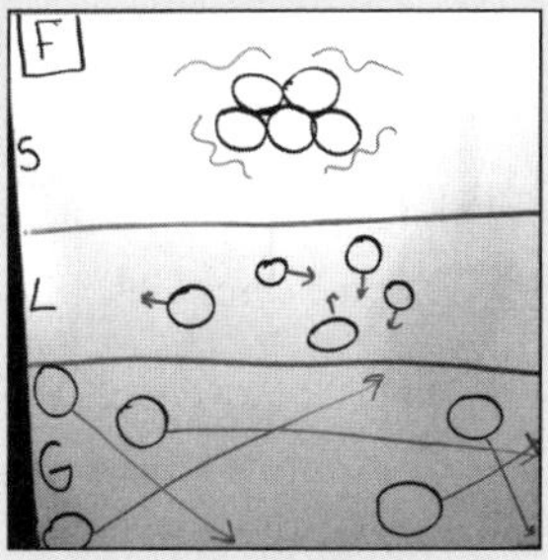
</td></tr>
</table>

Fig. 3.5. Representations created by each of the groups.
(Note that this is the second representation created by group F.)

165   The representations created by groups A and C (see fig. 3.5) raised similar issues. Although these groups did not change the size of the molecules as group B did, and group A showed no movement of molecules while group C did, both groups altered the spacing to show that molecules in the solid were the most tightly packed and those in the gas were the least tightly packed. In both cases

170   Ms. Nichols encouraged the groups to review what they had learned from experiment 2 and see how they could alter their representation to account for the pattern they identified in that experiment.

        When Ms. Nichols approached group F (Lei, Yvette, Stefan, and Marcellus), she noticed that they had made no progress on the task. Their draw-

175   ings (a picture of an ice cube, a puddle of water, and steam coming from a tea kettle) did not show what was happening at the molecular level. Ms. Nichols reminded the group that they were to depict a representation of the particles of water based on all the experiments together; a representation that explains how water molecules behave in all states of matter. Ms. Nichols engaged the students

180   in the following exchange:

Ms. Nichols: Okay, so let's look back at our experiments. What pattern did we notice in experiment 1?

Stefan: *(Reads from chart displayed at the front of the room)* "Solid water retains a definite shape and can't be compressed using a syringe. Liquid water flows to the

185   bottom of the container and is not measurably compressed in the syringe. Gaseous water escapes (disperses) out of the container and can be compressed when put into a syringe (when we push down on the plunger)."

Ms. Nichols: Right, so what does that mean? Can someone explain that in your own words?

Lei: Solid water is hard and can't be pressed when squeezed and liquid water is

190   moveable and the gas moves a lot.

Ms. Nichols: Okay, Stefan, does that make sense from what you read?

Stefan: Liquid water moves around and solid is hard and can't be squeezed. Yes.

Ms. Nichols: All right. And what about in experiment 2? Marcellus, tell us in your own words what pattern we noticed in experiment 2.

195   *Marcellus:* The volume of water at the freezing temperature changed, got bigger. The warmer vials of water all had the same volumes.

Ms. Nichols: Okay, Yvette. Do you agree with Marcellus' summary of the pattern we noticed from experiment 2?

Yvette: Yes.

200   *Ms. Nichols:* Yes? Okay, Yvette, I want you to summarize experiment 3 for your group just like we did for 1 and 2. Remember, we are looking for patterns. Then, drawing on these patterns, I want the group to create a representation that shows

what water looks like when we can see its particles with our magic glasses on. Don't forget that you have the marbles in your basket that you can use to help see how the molecules might interact. Okay?

205

Ms. Nichols hoped that the quick review of the patterns would help the students realize that each experiment gave some insight into the behavior of the water molecules. Before leaving the group, she suggested that they use the magnetic marbles in their baskets as a tangible representation to visualize the inter-

210 actions between water molecules. Her experience with group F confirmed her thinking that the task might be more complicated for some students. She made a note to emphasize (to other groups and in the future) that the four experiments they did and the patterns they noticed should help them in creating their representations for the Explaining the Behavior of Water task.

215 As Ms. Nichols continued to circulate around the room, she noticed group D's (Simone, Erik, Krystal, and Tatiana) representation. She had not anticipated students creating such an elaborate representation and was intrigued (see fig. 3.5). Although the group did not represent heat energy explicitly in the representation, they did show that ice floats in water and accurately portrayed the spacing

220 between molecules in different states. So, she engaged the group in a conversation to learn about their thinking.

Ms. Nichols:	Your representation is very detailed. You even show an ice cube floating! Could you tell me a little bit about why you depicted the water molecules in this way?

225 | *Simone:* | Okay, so at first we drew a whole different thing. We had the solid molecules all packed together and the gas ones spread out. |

| *Ms. Nichols:* | So why did you decide to change your representation? |

| *Simone*: | Well, Tatiana told us to. She said our drawing didn't make sense the way we had it. I don't know. |

230 | *Ms. Nichols:* | Tatiana, why did you say the original drawing didn't make sense? |

| *Tatiana:* | We saw how the water expanded when it froze. It took up more space, remember? *(referring to experiment 2)* |

| *Ms. Nichols:* | Yes. |

| *Tatiana:* | So we figured that the molecules have to spread out and push away from each
235 other when it's ice. That's why we showed the dotted lines here. *(points to drawing)* |

| *Krystal:* | Yeah, we drew the ice cube with the molecules of water held together with the dotted lines. See! *(points to drawing)* |

| *Ms. Nichols:* | Okay, why do the water molecules push away from each other in ice? What
240 do you think, Erik? |

Erik:	That's what we were just talking about. We know that it makes sense, but we aren't really sure why the molecules do that, like why they really spread out and push against each other when it's solid.

245

Ms. Nichols:	Okay, well, think about what might make water molecules push against each other when they are solid in ice. Don't forget that you have the marbles on your table to work with. Maybe they will help you make sense of what is going on here.

Ms. Nichols left the group with this parting thought, and the students continued to adjust their representation and manipulate the magnetic marbles.

250 　　As she approached group E (Kira, Nigel, Carlos, and Yuan) she noticed that like group D they had shown force enabling the solid to stay together but had not shown heat (see fig. 3.5). She asked Kira to explain their representation. Kira said, "Our representation shows how the solid takes up more space than the liquid. This is because there is something that actually pushes the molecules

255 apart further like a force. This is like when you take two magnetic marbles and hold them near each other. If you hold them so that these ends are near each other *(manipulating marbles with her hands)* you can feel the marbles being pushed apart. That's what is happening here." Ms. Nichols was impressed by her explanation of the role of force. As her parting question to the group, she asked

260 them to consider what caused water to transition from one state to another.

　　Before concluding her journey around the classroom, Ms. Nichols stopped in to see if group F had made any progress. She noticed that their new drawing (see fig. 3.5) showed molecules of the same size and that they indicated that the molecules were moving. While they were still missing some of the key features

265 of the explanation she was looking for, she felt that they had progressed far enough to be able to engage productively in the whole-class discussion.

　　At the end of twenty minutes, Ms. Nichols had completed the monitoring chart, as shown in figure 3.6. She was pleased to see that the groups had created a variety of representations and that she had done a good job in anticipating

270 what would occur. While there were a few surprises—the representation created by group D, the need to continue to take students back to the findings from the four experiments, and the fact that most students didn't use the marbles without prompting—she felt that she managed to handle things reasonably well. Armed with the data that she had collected, Ms. Nichols felt that she was

275 now able to determine which representations and features she wanted to focus on during the discussion.

STUDENT GROUP	Features Correctly Represented	Features Missing or Incorrectly Represented	ORDER and NOTES
A	__ SPACING __ HEAT __ MOVEMENT __ FORCES X_ TYPE & SIZE	X_ Incorrect Spacing X_ No/Incorrect Heat X_ No/Incorrect Movement __ Different-Size Molecules __ Other __ Accurate Representation	• Size of the molecules remains the same in all phases • Spacing in solid and liquid is inaccurate • Heat and movement of molecules isn't represented
B	__ SPACING X_ HEAT __ MOVEMENT __ FORCES __ TYPE & SIZE	X_ Incorrect Spacing __ No/Incorrect Heat X_ No/Incorrect Movement X_ Different-Size Molecules __ Other __ Accurate Representation	• Molecules get bigger from solid to gas instead of changing spacing of molecules; "molecules take up more space" • Heat is represented *Used marbles, but only after prompt
C	__ SPACING __ HEAT X_ MOVEMENT __ FORCES X_ TYPE & SIZE	X_ Incorrect Spacing X_ No/Incorrect Heat __ No/Incorrect Movement __ Different-Size Molecules __ Other __ Accurate Representation	• Movement represented – "solid squiggles" / arrows represent movement • Gas molecules are farther apart because you could squeeze the gas when the plunger went down
D	X_ SPACING __ HEAT X_ MOVEMENT X_ FORCES X_ TYPE & SIZE	__ Incorrect Spacing X_ No/Incorrect Heat __ No/Incorrect Movement __ Different-Size Molecules __ Other __ Accurate Representation	• Dotted lines show that solid takes up space in solid • Doesn't represent heat, but talked about freezing and water expanding in experiment • Shows ice floating in liquid water • Tatiana recognized ice expands when it freezes
E	X_ SPACING __ HEAT X_ MOVEMENT X_ FORCES X_ TYPE & SIZE	__ Incorrect Spacing X_ No/Incorrect Heat __ No/Incorrect Movement __ Different-Size Molecules __ Other __ Accurate Representation	• Force holds together solid • Talks about magnets and holding them to each other → alludes to idea of repulsion *Only group to use marbles spontaneously
F	__ SPACING __ HEAT X_ MOVEMENT __ FORCES X_ TYPE & SIZE	X_ Incorrect Spacing X_ No/Incorrect Heat __ No/Incorrect Movement __ Different-Size Molecules __ Other __ Accurate Representation	• Had difficulty understanding task at first – needed to review patterns! • Iconic representations only at first → similar to group C

Group A: Andre, Mack, Fatima, Bella • *Group B:* Kim, Kevin, Lamar, Rosa • *Group C:* Javier, Marques, Ree, Salima
Group D: Simone, Erik, Krystal, Tatiana • *Group E:* Kira, Nigel, Carlos, Yuan • *Group F:* Lei, Yvette, Stefan, Marcellus

Fig. 3.6. Ms. Nichols's completed chart for monitoring students' work during the
Explaining the Behavior of Water task

Analysis of Monitoring in the Case of Kendra Nichols

In part 2 of the Case of Kendra Nichols, we see a teacher who paid careful attention to what her students were doing during the lesson in an effort to document what they had done, support them in their work, and plan for a productive whole-class discussion. So what did Ms. Nichols actually do?

First, she collected data about what each group did, highlighting which features were correctly represented, which features were missing or incorrectly represented, and other notable aspects about each group's work, as shown in figure 3.6. The data made salient both the similarities among groups in the features they failed to include or represented incorrectly (e.g., four of the six groups showed incorrect spacing of molecules, five of the six groups did not include heat in their representations) and the differences in the extent to which their representations were complete and correct (e.g., groups D and E had representations that incorporated nearly all of the key features while groups A, B, C, and F either failed to include or misrepresented key features). The data paint a vivid picture of "where the class is" and will help Ms. Nichols in determining which representations and explanations to share and in what order (the next two practices). As Lampert (2001, p. 140) summarizes, "If I watch and listen during small group independent work, I am then able to use my observations to decide what and who to make focal" during whole-class discussion.

Second, Ms. Nichols took an active role in supporting students in making progress toward the learning goals of the lesson by helping students revise their representations when a feature was absent or had been misrepresented in some way. For example, she noticed that while group B had correctly shown the role of heat as causing the transition from solid to liquid to gas, they portrayed the molecules as getting larger with each transition. She encouraged students to review their findings from experiment 2 and consider whether their picture was consistent with their findings (lines 160–64). Rather than telling students what to do or hovering over them while they adjusted the representation, she gave them the opportunity to go back to the data they had collected from the experiment and reconsider their representation in light of their findings, and she left them to ponder this without her supervision. By leaving students with this "parting shot," the teacher gave them something specific to pursue and sent the message that they were capable of following through with the suggestion without her oversight. This in turn freed her to continue to make contact with other groups.

Third, Ms. Nichols pressed students who were on a correct pathway to think more deeply about what they were doing and what it meant. For example, while group D had a nearly correct and complete representation and explanation (they failed to account for the role of heat) not all students could explain the role of force. The teacher challenged them to consider why the water molecules push away from each other in ice (lines 239–40; 244–45) and encouraged them to use the magnetic marbles to make sense of what was going on (lines 245–47). Ms. Nichols wanted to make sure they could explain the representation they had created and was less concerned about adding heat to the representation since she felt that this feature would come out in the whole-group discussion.

Through her interactions with the groups, Ms. Nichols learned a great deal about her students' thinking. This information will help her in planning subsequent instruction, including, but not limited to, the discussion at the end of the lesson.

Conclusion

Anticipating and monitoring are crucial steps for teachers who want to make productive use of students' thinking during a lesson. By anticipating the key features that must be present for a complete and correct experimental design, explanation, or representation; the challenges that students are likely to encounter and/or the misconceptions that might surface; and the ways to support students as they work, a teacher is in a better position to recognize and understand what students actually do. Teachers who have engaged in this kind of anticipation and prediction can then use their understanding of student work to make instructional decisions that will advance the understanding of the class as a whole. Although a teacher can't anticipate everything that might occur in the classroom when a particular group of students engages with a specific task, whatever the teacher can predict in advance of the lesson will be helpful in making sense of students' thinking during the lesson. We see in the Case of Kendra Nichols that because Ms. Nichols had predicted much of what did occur, she was left with a limited number of "in the moment" decisions. But having taught the lesson once, Ms. Nichols now has a better sense of how students will respond, and she will be in an even better position to support learning the next time she teaches this lesson.

In this chapter, we saw a teacher who, as a result of her anticipating and monitoring, is ready to orchestrate a discussion of the Explaining the Behavior of Water task that builds on students' thinking. In the next chapter, we continue our discussion of the five practices with a focus on selecting, sequencing, and connecting. To do so, we will return to Ms. Nichols's classroom to see how the whole-group discussion unfolded.

TRY THIS!

Select a high-level task that has the potential to help students achieve a learning goal that you have identified. Individually, or in collaboration with one or more colleagues, do the following:

- Anticipate the key features that must be present for a complete and correct experimental design, explanation, or representation, as well as the challenges that students are likely to encounter and/or the misconceptions that might surface as they engage in the task.

- Consider questions that you could ask about the work that students are likely to produce that could help them in making progress on the task.

- Create a monitoring sheet that you can use to record data during the lesson.

CHAPTER 4

Making Decisions about the Discussion: Selecting, Sequencing, and Connecting

Once teachers have completed the work of monitoring—attending to what students are doing and saying as they work on a task, providing guidance as needed, and keeping track of who is doing what—they are ready to make decisions about the direction that the discussion will take. Central to the decision-making process is an awareness of the disciplinary ideas that they want their students to learn (as discussed in chapter 1) and what students currently know and understand related to those ideas (as reflected in the data collected by using the monitoring tool, as discussed in chapter 3). Teachers must then select which ideas and students to focus on to advance the understanding of the class as a whole, and they must sequence the sharing of students' work in such a way as to provide a coherent and compelling story line for the lesson. Finally, they must determine how they will connect these various approaches to one another and to the disciplinary ideas that are at the heart of the lesson. In this chapter, we present parts 3 and 4 of the Case of Kendra Nichols, considering how she used the data that she collected during the monitoring phase of the lesson to make decisions regarding the *selecting, sequencing*, and *connecting* of the representations produced by students. In considering these three practices, we first discuss practices 3 and 4, *selecting* and *sequencing*, together and then turn our attention to practice 5, *connecting*. In each section, we will begin by describing the practice or practices under consideration, and we will then present the relevant part of the Case of Kendra Nichols and analyze her use of the particular practice(s).

Selecting and Sequencing

Selecting is the process of determining which ideas *(what)* and students *(who)* the teacher will use to focus the discussion. This is a crucial decision, since it determines what ideas students will have the opportunity to grapple with and ultimately learn. Selecting can be thought of as the act of purposefully determining what core ideas students will have access to—beyond what they were able to consider individually or in small groups—in building their understanding of the discipline. Selecting is critical because it gives the teacher control over what the whole class will discuss, ensuring that the science concepts and ideas that are at the heart of the lesson actually get on the table. We have come to think of the question, "Who wants to present next?" as either the bravest or most naïve invitation that can be issued in the classroom. By asking for volunteers to present, teachers relinquish control over the conversation and leave themselves—and their students—at the mercy of the student who has volunteered to take center stage. Although this may work out fine—what the student presents may be both understandable and connected to the lesson goal—unfiltered student contributions can be difficult to follow, take the conversation in an unproductive direction, or shut down further discussion.

Consider, for example, the iconic representation (i.e., pictures of an ice cube, a puddle of water, and stream coming from a tea kettle) that group F (in Ms. Nichols's class) initially created in response to the Explaining the Behavior of Water task. If the teacher had simply asked, "What group wants to go next?" and group F volunteered to share their initial drawing with the class, much time could have been spent discussing a representation that was unlikely to advance students' understanding about the molecular behavior of water. On the other hand, unwittingly calling on a group that has produced a sophisticated response too early in the discussion can shut down further discussion. It can also make those students with less complete or correct work feel unsure of themselves and unable to see how their thinking relates to a more finished or polished product. For example, recall that group D in Ms. Nichols's class drew a more sophisticated representation of the behavior of water where they detailed the movement of the water molecules in all three phases, the molecular structure of the solid molecules, and the ability of gas molecules to move beyond the space of a container (see fig. 3.5). If Ms. Nichols had called on group D to explain their drawing to the class first, these more complex ideas would have been introduced before many other students' misconceptions and incomplete ideas were addressed. Groups A, B, C, and F then might not initially have been able to understand or connect to group D's thinking, leaving them with an incomplete understanding of the molecular behavior of water.

Although selecting is first and foremost about what core ideas in science will be highlighted, it is also about who will present them. For example, in Mr. Gates's class, three groups had used line graphs to show the growth of a typical Fast Plant. If Mr. Gates determined that this is a strategy that he wants his students to consider (the *what*), he will then need to determine which student he will ask to share with the whole class (the *who*). In making this decision, he may want to consider which student has not presented recently and give that student an opportunity to take center stage in the classroom. Mr. Gates employed this strategy in the case of the Fast Plant task when he decided to have Ryanne from group 7 share her group's work with the class because she had not presented in several days. A periodic review of completed monitoring sheets collected from previous lessons would provide a record of which students shared their work in the recent past and be a useful way for teachers to track students' contributions over time. This will ensure that all students have the opportunity to be seen as authors of ideas and to demonstrate their competence (Lotan 2003).

Sequencing is the process of determining the order in which the students will present their work. The key is to order the work in such a way as to make the ideas accessible to all students and to build a coherent story line. For example, if Bri, a student who was working on the Fast Plants task in Mr. Gates's class, had presented her group's work (see fig. 2.6) first instead of last, it might have been challenging for students to understand, since the approach her group took—unlike that of any of the others—featured a box-and-whiskers plot which made salient the low and high values, the median, and the values that separated the top quarter and bottom quarter. Instead, Mr. Gates decided to have group 7 present first, as several groups had used this strategy and it would therefore, the teacher reasoned, be more accessible to the entire class.

Presenting the most commonly used strategy first is one approach to sequencing student work, but it may not always be the best way to proceed. For example, if a misconception surfaces during work on a task, the teacher may want to begin the discussion by addressing this issue directly. Another alternative is for the teacher to sequence solutions that move from less to more sophisticated or complete. Consider, for example, the genotype task that appeared on the National Assessment of Educational Progress in 1996 (fig. 4.1). Although response 1 is the most complete response, since it addresses the genotype of the father as well as indicating that the genotype of the father's parents would be needed to prove the genotype of the father, it might not be the best starting point. Rather, the teacher might want to start with a less complete response, such as response 3, and use it to generate a discussion about dominance and what one would need to know to conclude that the allele carried by the father is dominant. Attention could then turn to response 2, where students could discuss both what additional information is provided and what information is still needed. The discussion could conclude with a focus on response 1, highlighting why the genotype of the father can only be confidently determined by learning the genotype of his parents. This sequence would bring all students into the discussion, as the less complete response would be accessible to everyone. Each successive strategy could be carefully tied to those that came before it so that students could ultimately see how the most complete response is related to the less complete approaches. It is worth noting that in this scenario the teacher did not ask the students who produced response 4 to share their work publicly. Given the limited amount of instructional time available, it is not reasonable to have every group present every day. The teacher must make choices about what responses are likely to advance the understanding of the entire class. He or she may want to make a point to check in with groups who have incorrect or incomplete responses that were not discussed in order to make sure that by the end of the discussion they understand that limitations of their own work.

We now return to the Case of Kendra Nichols, picking up this time where Ms. Nichols is in the process of planning the *what* and the *who* for the whole-class discussion of the Explaining the Behavior of Water task.

ACTIVE ENGAGEMENT 4.1

- Review Ms. Nichols's completed sheet for monitoring her students' work, as shown in figure 3.6.
- Given Ms. Nichols' goals for the lesson (see the Case of Kendra Nichols—Part 1 in chapter 3, lines 5–24), determine which representation you would want to have shared, and in what order, during the discussion portion of the lesson, in order to accomplish the stated goals.

Task	A mother with attached earlobes and a father with free earlobes have 5 children— 4 boys and 1 girl. All of the children have the father's type of earlobes. What can be predicted about the genotype of the father? Construct a genetic diagram to support your prediction. What additional information, if any, would you need to determine the genotype of the father? Explain.

Response 1	Response 2	Response 3	Response 4
The genotype of the father is homozygous. The genotypes of the parents of the father would be needed to prove this, however, because the father inherits his traits from his parents.	The father has the dominant gene for ear lobes while the mother has the recessive gene.	The father's genes are more dominant over the mother's.	The bloodline comes through the father so maybe in some way that is connected with why their earlobes are the same as the father's.

Response 1:

Father   **Mother**
FF   **ff**

Ff   Ff   Ff   Ff   Ff

F = free earlobes
f = attached earlobes

Response 2:

Father EE

	E	E
e	Ee	Ee
e	Ee	Ee

Mother ee

Therefore all the children would have both dominant and recessive genes for the attachment of the earlobes, but the dominant gene is the one that is shown.

Fig. 4.1. The genotype task (O'Sullivan and Weiss 1999, pp. 171–73)

Matter and Molecules: The Case of Kendra Nichols
(Part 3—Selecting and Sequencing)

Ms. Nichols was now ready to make decisions about which representations to have presented during the discussion and the order in which they should be presented. Based on her learning goals for the lesson (see chapter 3, the Case of Kendra Nichols—Part 1, lines 5–25), and what she had learned about students' conceptions while monitoring their work, she decided that the whole class discussion would target the following features of water:

280

1. Molecule spacing changes from one phase to another, but molecules are most closely packed in the liquid phase and farthest apart in the gas phase;

285

2. The type and size of the water molecule stays the same in all phases;

3. Solid molecules are connected by attractive forces, which hold the molecules in a crystal structure; when motion increases, the relative effect of the forces decreases;

4. The motion of molecules changes from phase to phase, with the most motion (in terms of distance and direction) happening in the gas phase and the least in the solid phase;

290

5. Gas molecules spread out beyond the container of water while the molecules of ice and water are "contained"—either self-contained (solid ice) or

contained by a physical boundary (the glass holding water); and

6. Increasing heat energy results in changes of phase (e.g., from liquid to gas) and decreasing heat energy results in changes of phase (e.g., from gas to liquid).

With these ideas in mind, she reviewed the monitoring chart that she had completed while she observed and interacted with the groups (fig. 3.6) and began to make decisions about what features should be presented, who should present them, and in what order they should be presented. She wanted to start the discussion by exploring the feature of molecule spacing since four of the six groups (see solutions A, B, C, and F in fig. 4.2) drew representations with incorrect spacing between the solid and liquid phases. She decided to focus on the representation produced by group A. Ms. Nichols wanted to point out a key "right" idea about consistency of molecular structure—that the molecules are the same size in all three phases. In addition, she would be able to point out that many groups had similar spacing between molecules in different phases, with the least amount of space between molecules in ice (solid) and the most amount of space between molecules in vapor (gas). This would allow the discussion to focus on the feature she wanted to highlight (spacing) before addressing other features.

Ms. Nichols planned to move to group D next and ask them to address how they depicted the spacing of their molecules and why they drew the lines in the solid phase. This would focus students' attention on two areas where group A's representation fell short. In doing so, she hoped to launch a discussion about the feature of attractive forces between the water molecules where she would introduce the analogy that water molecules act like magnets. To help students understand the analogy, she would turn their attention to the magnetic marbles in order to experience attractive and repulsive forces between the magnets. She would then address the way group D depicted the gas molecules, which are not only in the container, but in the space outside the container as well.

Next she decided to have group E explain their thoughts about the motion and speed at which the molecules move in all phases. Ms. Nichols wanted students to notice that the forces between the molecules decreases as their motion increases. Prompting students to hold the magnets in various ways, she hoped they would feel how the amount of force changes in relationship to the distance between the molecules and the speed at which they are moving.

She planned to end the discussion with the representation created by group B, since it was the only one that addressed the feature that heat energy results in the changes in state of matter. She wanted the students in the class to consider how heat energy influences the molecules' speed and movement and the importance of heat energy in this process. In addition, Ms. Nichols wanted to check in with group B to see if they wanted to reconsider the way they depicted their molecules (as changing in size and not changing in spacing).

335   As Ms. Nichols made her decisions, she added a few notes to her monitor-
       ing chart (fig. 4.3, which begins on the facing page), to remind herself of the
       points she wanted to make and features she wanted to highlight. She also noted
       that none of the groups created a representation depicting all the features, so she
       wanted to make sure that the students would have time to modify their repre-
340   sentations based on the features discussed. In the interest of time, Ms. Nichols
       did not want all the groups to present, particularly groups that had similar
       representations. She knew from past experience that this took too much time,
       added little to the learning opportunities, and often led to students' off-task
       behavior and waning interest.

Group	Representation Drawn	Group	Representation Drawn
A		C	
B		F	

Fig. 4.2. The four student representations with incorrect spacing of molecules

STUDENT GROUP	Features Correctly Represented	Features Missing or Incorrectly Represented	ORDER and NOTES
A	__ SPACING __ HEAT __ MOVEMENT __ FORCES X_ TYPE & SIZE	X_ Incorrect Spacing X_ No/Incorrect Heat X_ No/Incorrect Movement __ Different-Size Molecules __ Other __ Accurate Representation	• *Size of the molecules remains the same in all phases* • *Spacing in solid and liquid is inaccurate* • *Heat and movement of molecules isn't represented* *MOLECULE SPACING changes from one phase to another. *The TYPE and SIZE of the molecules stays the same. • #1 Fatima
B	__ SPACING X_ HEAT __ MOVEMENT __ FORCES __ TYPE & SIZE	X_ Incorrect Spacing __ No/Incorrect Heat X_ No/Incorrect Movement X_ Different-Size Molecules __ Other __ Accurate Representation	• *Molecules get bigger from solid to gas instead of changing spacing of molecules, "molecules take up more space"* • *Heat is represented* *Used marbles, but only after prompt *Increasing HEAT ENERGY results in changes of phase (e.g. from liquid to gas) and decreasing heat energy results in changes of phase (e.g. from liquid to solid). *How much a molecule is moving is due to the amount of heat energy it possesses. So the more heat energy, the faster the molecules move on average and the farther apart (on average) they are, etc. #4 Lamar
C	__ SPACING __ HEAT X_ MOVEMENT __ FORCES X_ TYPE & SIZE	X_ Incorrect Spacing X_ No/Incorrect Heat __ No/Incorrect Movement __ Different-Size Molecules __ Other __ Accurate Representation	• *Movement represented – "solid squiggles" / arrows represent movement* • *Gas molecules are farther apart because you could squeeze the gas when the plunger went down*
D	X_ SPACING __ HEAT X_ MOVEMENT X_ FORCES X_ TYPE & SIZE *MOLECULE SPACING changes, but we see that they are most spaced out in the gas phase and most closely packed in the liquid phase.	__ Incorrect Spacing X_ No/Incorrect Heat __ No/Incorrect Movement __ Different-Size Molecules __ Other __ Accurate Representation	• *Dotted lines show that solid takes up space in solid* • *Doesn't represent heat, but talked about freezing and water expanding in experiment* • *Shows ice floating in liquid water* • *Tatiana recognized ice expands when it freezes* *Gas molecules spread out beyond the container of water while the molecules of ice

STUDENT GROUP	Features Correctly Represented	Features Missing or Incorrectly Represented	ORDER and NOTES
D	*Solid molecules are connected to one another in some way (FORCES).		and water are "contained"---either self-contained (solid ice) or contained by a physical boundary (the glass holding water). #2 Krystal
E	X_ SPACING __ HEAT X_ MOVEMENT X_ FORCES X_ TYPE & SIZE	__ Incorrect Spacing X_ No/Incorrect Heat __ No/Incorrect Movement __ Different-Size Molecules __ Other __ Accurate Representation	• Force holds together solid • Talks about magnets and holding them to each other → alludes to idea of repulsion *Only group to use marbles spontaneously *MOTION of molecules changes from phase to phase, with the most motion (in terms of distance and direction) happening in the gas phase and the least in the solid phase. *When motion increases, the relative effect of the attractive forces decreases. #3 Yuan
F	__ SPACING __ HEAT X_ MOVEMENT __ FORCES X_ TYPE & SIZE	X_ Incorrect Spacing X_ No/Incorrect Heat __ No/Incorrect Movement __ Different-Size Molecules __ Other __ Accurate Representation	• Had difficulty understanding task at first – needed to review patterns! • Iconic representations only at first → similar to group C

Group A: Andre, Mack, Fatima, Bella • Group B: Kim, Kevin, Lamar, Rosa • Group C: Javier, Marques, Ree, Salima
Group D: Simone, Erik, Krystal, Tatiana • Group E: Kira, Nigel, Carlos, Yuan • Group F: Lei, Yvette, Stefan, Marcellus

*Ideas to emphasize!!

Fig. 4.3. Ms. Nichols's completed chart for monitoring students' work on the Explaining the Behavior of Water task

345   Once Ms. Nichols had decided which groups would present, she needed to determine which student would be the group's primary spokesperson. Although she sometimes had the entire group make the presentation, this strategy often resulted in one student doing the majority of the talking and the others becoming part of the background. She reviewed the membership of the tar-
350   geted groups and identified presenters who did not have a chance to share their work in the last week (as shown in the far right column of the chart in fig. 4.3). Because Ms. Nichols previously established the expectation that any member of a group could be asked to present, every student needed to understand the work that their group produced well enough to discuss it in front of the class. She
355   found that this expectation helped her hold all students accountable for participating in small-group collaborations.

Analysis of Selecting and Sequencing in the Case of Kendra Nichols

In part 3 of the case, we see a teacher who thoughtfully considered how to use the work produced by her students as a basis for a whole-class discussion. She wanted to discuss features of water in solid, liquid, and gaseous form that included molecular size, spacing, force, motion, and the role of heat energy, consistent with her learning goals (see part 1 of the Case of Kendra Nichols, chapter 3, lines 5–25). No group had accurately depicted all of the features of water, but through her careful monitoring Ms. Nichols identified aspects of representations that would allow her to highlight particular features. She decided to start with group A's poster so she could point out the correct idea about the size of molecules, then move to group D's representation to discuss molecular spacing and force, followed by group E's work and a discussion of motion, and ending with group B's work and a discussion of the role of heat.

This sequencing started with a representation (group A) that had only one correct element (molecule size) and a common misconception (the molecules in ice were most tightly packed), featured two strong representations in the middle (groups D and E), and ended with a representation (group B) that again, had only one correct element. Ms. Nichols positioned the students to consider the new element (heat) and its role in transitioning from one phase to another, and to consider the limitations of group B's drawing given the discussions of the representations produced by D and E.

The monitoring sheet that Ms. Nichols created and used during the lesson (see fig. 4.3) made it possible for her to keep track of the features correctly represented by each group as well as those missing or incorrectly represented and to make note of the aspects of the representations she wanted to highlight. By closely attending to students' work, she was able to identify features that would be likely to provoke the thinking of the entire class. Hence, she used her goals for the lesson and her knowledge of "where various students' thinking was" as they worked on the tasks in groups to guide her decisions regarding what ideas would be put on the table for the discussion and in what order.

From her careful monitoring, Ms. Nichols knew that four of the six groups did not represent molecular spacing correctly and therefore wanted to make sure that all students understood why the molecules in ice are spaced further apart than the molecules in water. To accomplish this, she decided to use the magnetic marbles so that students could experience attractive and repulsive forces. Also, she noted on her monitoring sheet that groups B and E used the marbles so she would be able to invite them to discuss their experiences with the marbles.

Although it is clear that Ms. Nichols made thoughtful decisions about the best ways to highlight the ideas to be learned, she also made a point to invite students to present who had not shared their work in the past week (lines 345–51). By selecting students who had not presented recently, she was giving them the opportunity to demonstrate their competence and to gain confidence in their abilities. Her practice of identifying one member of the group to present was also a way to hold all members accountable for the work of the group.

There are many different ways student responses could be selected and sequenced that could be equally productive. The point is that the method selected must support the story line that the teacher envisions for the lesson so that the desired science ideas and concepts emerge in a clear and explicit way. Ms. Nichols's work suggests that she was well positioned to orchestrate such a discussion.

Connecting

Connecting may in fact be the most challenging of all of the five practices, because it calls on the teacher to craft questions that will make the core disciplinary ideas visible and understandable to the entire class. The questions must go beyond merely clarifying and probing what individual students did and how they did it. Instead, they must focus on meaning and relationships and make links between ideas and representations. It is also important to pose questions that engage all members of the class, not only the presenters.

Although questions need to expose the ideas to be learned in an explicit way, they must begin with what students know. Moving between where a learner is and where one ultimately wants him or her to end up "is a continuous reconstruction" (Dewey 1902, p. 11). This notion suggests that the teacher needs to know both the ideas to be learned and what students know about these ideas in order to bridge the two worlds. To consider one without the other can result in questions that draw blank stares from students because they make no connection with their current ways of thinking. To consider one without the other can also result in students' thinking remaining stagnant, instead of moving toward new disciplinary understandings. For this reason, framing questions in the context of students' work is critical.

For example, imagine that a teacher's goals for a lesson built around the genotype task (fig. 4.1) are for students to know that (1) a genotype for a particular monogenic trait is composed of two alleles—one from each parent; (2) if one allele is dominant and the other recessive, a heterozygous organism will display the dominant trait (phenotype); and (3) genotype can only be determined conclusively by knowing the genotype of both parents. In a discussion of the solutions to this task (fig. 4.1), the teacher might want to ask students how the claims made in response 3 ("The father's genes are more dominant over the mother's") and response 4 ("The bloodline comes through the father so maybe in some way that is connected with why their earlobes are the same as the father's") can be supported or refuted by the diagrams and explanations provided in responses 1 and 2. Through this discussion, students may come to see that the allele for free earlobes is likely dominant given the phenotypes of the five children (all free earlobes), and therefore the genotype of the father could be FF or Ff. Both options could be explored and students would likely conclude that FF is more likely, but it is necessary to know the genotypes of the father's parents to confirm this idea.

Now suppose that while anticipating students' responses to this task, the teacher thought it unlikely that her students would propose a solution in which the father carried the heterozygous (Ff) genotype. Knowing that this is an important possibility for them to consider, the teacher created a fictitious student response in which the father carried this Ff genotype (fig. 4.4). She kept this work ready to present to the students during the discussion as "work from another class." In doing so, the teacher was positioned to provide students with an opportunity to engage in further discussion about the underlying genetic concept of monogenic inheritance.

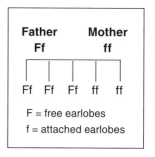

Fig. 4.4. Teacher's anticipated "work from another class" for the genotype task

The key to connecting is making sure that the teacher openly addresses the ideas to be learned. Consider, for example, the Case of Nathan Gates presented in chapter 2. Mr. Gates pressed students to "see" if they could describe the phenomenon articulated by Tess (lines 112–13) that resulted in the identification of the basic S shape and the realization that the shape was not dependent on the height of the plant. In addition, through the analysis of several graphs, Mr. Gates was able to highlight the point that Bri made, namely that showing a range of data is important (lines 224–25). By questioning students about the carefully sequenced work, Mr. Gates was able to help them understand two key ideas—that typical growth in Fast Plants is described by range and shape and that Fast Plant growth is characterized by an S-shaped growth curve. While Mr. Gates did not focus on making connections between the groups' representations beyond pointing out that the line graphs produced by groups 2, 3, and 7 all had the S-shape, there are other connections he could have made. Specifically, he could have asked group 8 how their bar graph compared to the line graph presented by group 7. This question might have surfaced the fact that while group 7 showed all the plant heights for each day, group 8 presented the average of the heights. Group 8 might have also pointed out that if they connected the midpoints of their bars into a line graph, then they too would have an S-shaped curve. Similarly, students might have tried to find connections between the final graph presented by group 5 (box-and-whiskers) and the graphs produced by groups 7 and 8. Like group 7, group 5 showed the range of values on a given day, and like group 8, they showed a measure of central tendency but selected the median instead of the mean.

Important for facilitating the actual discussion is considering how the students will share the work. For example, Mr. Gates had his students create posters of their work because he thought that it would be easier to look across different representations during the whole-class discussion. (It is worth noting that the expense of chart paper may limit its use; in addition, making posters can turn science activities into time-consuming art projects, and posters will be less effective if all the work is likely to be similar.) Ms. Nichols had students draw their representations for the Explaining the Behavior of Water task on whiteboards. This way, students could modify their drawings as needed without having to use additional material resources. Alternatively, document projectors that allow students to display their original work without having to recopy it on a poster or transparency may be effective and efficient. With the document camera, the teacher can show more than one solution by using the zoom to make the images smaller.

We now return to the Case of Kendra Nichols to see how Ms. Nichols helped her students make connections during the discussion.

ACTIVE ENGAGEMENT 4.2
- Review Ms. Nichols's completed monitoring chart shown in figure 4.3.
- Given Ms. Nichols's goals for the lesson (see the Case of Kendra Nichols—Part 1 in chapter 3, lines 5–25), identify questions that you would want to ask students about the representations to achieve these goals.

Matter and Molecules: The Case of Kendra Nichols
(Part 4—Connecting)

Ms. Nichols began the final phase of the lesson (sharing and summarizing) by placing the six boards across the front of the room in the following sequence: group A, group D, group E, group B, group C, and group F. Although she didn't plan to discuss the work of groups C and F formally, she intended to invite these groups to connect what they had done to the presented work.

360

Ms. Nichols started the discussion by asking Fatima from group A to come up and explain her group's representation.

Fatima: We show how solids are packed together and liquids are not as packed. But the gases are spaced out. We were thinking about how the gas in the plunger—the syringe thing—was able to be compressed. We put it in there and squeezed and the plunger went down. So we think that's because the molecules have all this space or air between them and when you compress them they are moving closer together. But, like a solid you can't push them closer together because they're already as close as possible. That's why solids keep their shape—you can't move the molecules around. We weren't really sure about the liquid molecules. We couldn't compress them—or compress the liquid in the syringe. But we think they're not stuck together like in a solid.

365

370

Ms. Nichols: So you are saying that all three of these phases of water are made up of the same kinds of molecules but what is changing is the way they are spaced, how close or far apart they are from one another? Is that right?

375

Fatima: Yes, it's all water molecules. They are all the same, water never changes, just they are spaced differently in each phase.

Ms. Nichols: Okay, does anyone in Fatima's group have anything to add? Bella, does her explanation describe what you discussed in your group?

380

Bella: Yup. She covered it.

Ms. Nichols knew that the work of many of the other groups showed a similar relationship in terms of molecule spacing and thought that the issue of spacing could be addressed through a discussion of group D's representation. Towards that end, Ms. Nichols commented, "I noticed that group D changed the spacing from phase to phase, but it's not quite the same as in group A." She invited Krystal (from group D) to talk about how her group was thinking about the spacing of the molecules. The following exchange occurred:

385

Krystal: Yeah, well, at first we drew something totally different. We drew a picture like group A's at first.

390

Ms. Nichols: All right. So you also had a picture like A's at first. What made you change your drawing?

395	*Krystal:*	Well, we remembered that the frozen water took up more space in our experiment, so we changed the drawing and put more space between them. So, we figured the molecules have to spread out and push away from each other *(points to the dotted lines)*. And the gas is moving all around and spaced really far apart.

400 Ms. Nichols explained that she liked how group D used the patterns they noticed in the experiments to draw their representation. However, she realized that many other groups were struggling with molecule spacing, and she wanted to address this idea in more depth before moving on with the discussion to ensure that all the groups understood the spacing relationship. She said, "Okay, we see that other groups have representations that show the same relationship
405 as group A in terms of how the molecules are spaced. Now, I want you to do a thought experiment with me. Assume that group A's representation is accurate, and that the molecules in ice are closer together than the molecules in water. Okay, now imagine that these marbles that I have here are molecules of water. *(Shows students a glass jar filled with glass marbles.)* Imagine that I could push
410 them together even closer than they already are in this jar. So, if I could push them closer together, pack them tighter, would I be able to put more or less of them in this particular jar? Take 30 seconds and think about your answer." She walked around the room showing students the jar of marbles finally calling on Javier (from group C) to answer.

415	*Javier:*	There would be more marbles in the jar.
	Ms. Nichols:	Right, there would be more marbles that fit in the jar. Good. So, if you think about the marbles as molecules, there would be more molecules in the same space because I am not changing the size of the jar. Okay, now, imagine I have a beaker of water and the molecules are spaced like they are in group
420		A's picture. And I have an ice cube with molecules spaced like this *(points to group A's whiteboard)*, closer together. Would there be more water molecules in the same amount of ice compared to water? Would you expect the ice to take up more space than the liquid water?
	Kim:	If you had more molecules in the ice, the ice would take up less space.
425	*Ms. Nichols:*	Okay, class. Kim says that the ice would take up less space. Do you agree or disagree? Andre, do you agree or disagree with Kim?
	Andre:	Agree.
	Ms. Nichols:	Andre, tell me more. Why do you agree with Kim?
	Andre:	Well, if you squish the water marbles in the jar smaller like in A so it becomes
430		ice, it would be smaller because there are more molecules packed in such a small space.
	Ms. Nichols:	Right, we'd expect the ice to take up less space than water because the ice

		would be packed more tightly than water. But what happened when we put ice in our water?
435	*Students:*	It floats!
	Ms. Nichols:	Hmmm. So, ice floats. Why do you think the ice might float? Salima *(from group C),* what do you think?
440	*Salima:*	Well, frozen ice floats so it can't be heavier than water. If it's lighter and floats it has to have less molecules. I remember the experiments we did. Ice takes up more space than water.
	Marques:	*(Interrupts)* Ice floats so it can't have more molecules, it has to have less. There has to be something holding the ice molecules apart like that so they don't get close.
445	*Ms. Nichols:*	All right. So I want to draw our attention to something here. We are not really talking about the molecules being heavier or lighter. We are talking about the amount of molecules in a given space. When we move into this idea of heavy and light, we are really talking about density, which is something we will address later. Let's remember to focus on the idea that ice has fewer molecules in a given space than water. Does that make sense for everyone?
450	*Students:*	Okay.
	Ms. Nichols:	Okay, so, Krystal, I'm wondering about the lines that you have drawn between the molecules in ice. Could you say more about those?
455	*Krystal:*	Okay, well, we have lines between the molecules. We think that the molecules have to push away from each other somehow because they are farther apart than the water. So whatever is pushing them apart, kind of holds them in place, too, because it is solid. Let's see. Like those DNA molecules Mr. Jones had in Life Science class. You know, the molecules are all twisted and those sticks hold them together.
460	*Ms. Nichols:*	So you are saying that there is some kind of force or connection that holds the molecules in place? That is why ice keeps its shape? What do you all think about this idea?
	Ree:	I can remember those DNA molecules Mr. Jones showed us in Life Science. The molecules have to be held together by something. Even though ice isn't in our bodies, it's still molecules.
465	*Ms. Nichols:*	So does the idea of forces make sense to you?
	Lei:	That makes sense, forces hold the ice together. Cool.
	Ms. Nichols:	Now I see that you decided to make the molecules in ice farther apart because that would explain why it takes up more space than when it was just liquid water. And I see that you decided that connections—can we call these

470   "forces?"—would explain why the molecules were locked into place. That's a pretty powerful idea. But I wonder if anyone thought about why the forces would actually hold the molecules a little farther apart than in water, as opposed to pulling them even closer together. And I have a tool that I want to share with you to help you think about that.

475   At this point, Ms. Nichols took a moment to write the spacing feature on the board for students (see row 1 in fig. 4.5). She wanted to keep track of the important features of water that emerged during the discussion so that students could use the information to modify their representations at the end of class. Ms. Nichols then explained that some groups (groups D and E) used the mag-
480   netic marbles to help them think about the behavior of the water molecules. She asked these groups if they had anything to add about the forces between the marbles. Nigel (from group E) raised his hand.

Nigel: When we played with the marbles, we could feel them being pulled together when they were close.

485   Ms. Nichols: So let me see if I understand. When you hold the marbles close together, you can feel the attraction between the marbles. Am I understanding you correctly?

Nigel: (Nodding head) Yes.

Erik: (Interrupts) And if you hold them way out here (holding marbles far apart),
490   you don't feel it. You don't feel the magnet.

Ms. Nichols: Okay, when you hold the marbles farther apart you don't feel the attraction, right? I would like everyone to try something for me.

	Feature of Water	Solid (Ice)	Liquid (Water)	Gas (Vapor)
1	Spacing	Farther apart than liquid	Closest	Far apart
2	Forces	Present – holds ice structure – noticeable	Not as strong	Not important
3	Motion	Vibrating	More movement	Fast movement
4	Heat Energy	Add heat→	Add heat → ← Remove heat	← Remove heat

Fig. 4.5. Summary of the features of the behavior of water that Ms. Nichols created during the class discussion

She quickly distributed magnetic marbles to each group and asked the stu-
495   dents to take a few minutes to see if they could notice the attraction between the magnets that Nigel and Erik described. After they took some time to manipulate the marbles and notice the forces, Ms. Nichols showed the class a diagram of water molecules so students could see how the hydrogen and oxygen

molecules are aligned. She showed figure 4.6 to the class and explained that water molecules are held in a lattice structure so that the hydrogen atoms are

500   close to the oxygen atoms, and the oxygen atoms aren't close to each other. She noted that they now had a representation showing that the molecules of water are spaced differently in each phase and that the molecules in ice are held in a stable structure because of their forces; she then recorded the new feature on the board (see row 2 in fig. 4.5). Before moving on to the next idea, Ms. Nichols

505   drew the students' attention to the gas molecules and how group D depicted the molecules as distributed "all over the place." In order to bring in the feature of motion, she decided to move on to group E.

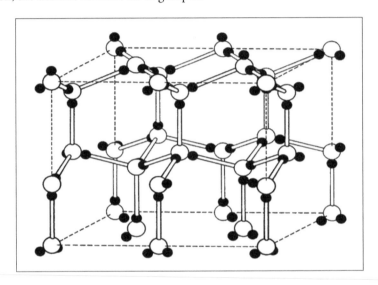

Fig. 4.6. Image of crystalline structure of ice Ms. Nichols used during discussion. (From *The Chemical Physics of Ice* by N.H. Fletcher. Copyright © 1970 Cambridge University Press. Reprinted with the permission of Cambridge University Press.)

Ms. Nichols then asked Yuan (group E) to share her group's ideas about motion with the class. Yuan explained, "Our representation shows why the

510   gas molecules can move all around outside space or even inside space, too, like group D did. In the solid state, the molecules are vibrating *(points to solid's vibration lines)*. The liquid moves a little more, but not as much as the gas. Gases move everywhere, no specific places. Liquids move less so they can take up the space of the container they're in, like their drawing *(points to group D's represen-*

515   *tation)*. Oh, and the solid is just vibrating and we drew lines too because they are held together and are like really pressed apart a little, that's why solids take up more space than liquid."

Ms. Nichols thanked Yuan for her explanation and said, "So if we add the idea of motion to our representation, we can explain why gas molecules are

520   traveling freely around a room while water and solid molecules are confined. Gas molecules move in all directions and fast. Solid molecules are locked into a structure, so they can just vibrate or wiggle. And water molecules can move

525

530

past each other but not as much as in a gas." She then prompted the students to think about the forces introduced by group D. She asked the class if the liquid and gas phases have forces between the molecules just like the solid phase. Once again, the students used the magnetic marbles to feel the force when the molecules were close together. As she told the students to move the marbles representing the molecules apart at various speeds, the students discussed what they felt. Ms. Nichols explained that just like the magnetic marbles, when water molecules are moving fast enough and are far enough away from each other, these forces are not as important. Once the molecules slow down and get closer the forces are more noticeable between the molecules.

535

Before moving on, Ms. Nichols asked the class if anyone had any questions about the features discussed so far. Since the students did not have any questions at this point, she marked the motion feature on the board (see row 3 in fig. 4.5). She then explained that she wanted to discuss the molecular representation that group B produced. She asked Lamar from group B to tell the class why his group included heat in their model. The following short exchange occurred:

540

Lamar: Well, our representation's a little different. I guess. We drew the molecules wrong, but in our drawing, they change when you add heat. See *(points to representation)* when you add heat to ice, you get water, and when you add heat to water, you get gas. So you gotta take away heat to go from gas to liquid to solid. I don't know how you really take away heat, but you do. That's it.

545

Ms. Nichols: Okay, in your picture, are your molecules moving?

Lamar: No, they aren't moving. We thought they changed size to get to the different phases by adding heat or taking away heat. But it makes sense that the molecules stay the same size.

Ms. Nichols: All right, Lamar. So you drew here that the addition of heat to water, for example, will result in its becoming a gas. And if you connect that with group E's representation, what can you say is happening to the molecules? Lei?

550

Lei: Well . . . if you add heat to ice it melts.

Bella: I know what you mean. Adding heat to the ice makes the molecules move more so they move apart, right? Then adding heat to water creates the gas.

555

Ms. Nichols: Carlos, what do you think about what Lei, Lamar, and Bella said? Do you agree or disagree?

Carlos: Uh, well. I agree.

Ms. Nichols: Okay. Why do you agree, Carlos?

Carlos: Yeah. Well, my mom makes tea all the time. She boils the water in the kettle on the stove. So she adds the heat and the water boils and some of it becomes

560

steam and you can see it shooting out the top—that's the gas. You can put that water in the freezer and make it so cold it would freeze. So to get the water to boil you add heat and the molecules move faster and farther and farther until they are a gas.

565   *Ms. Nichols:*   Right! I really like your example, Carlos. They are moving more and getting farther away. So when we are thinking about molecules, we can think about heat energy as energy of motion. Increasing heat energy makes the molecules move more. And taking away heat energy, cooling something down, will result in less motion of the molecules. Let's add this to our representation.

570   Ms. Nichols added the idea of heat energy to the class representation (see row 4 in fig. 4.5). Before closing the lesson, she wanted to check group B's understanding of the molecules' size and spacing. After making certain that this group realized the molecules do not change size and grasped how the spacing varies between the phases, she was sure that the students had an understanding
575   of the behavior of water molecules. Before moving on, Ms. Nichols wanted to check in with groups C and F because their representations were not the focus of the discussion during the lesson. After asking each group if they had anything to add to what had been said, Stefan (from group F) responded, "I don't have anything to add. Are we going to get to fix our drawings?" Ms. Nichols thought
580   this was an excellent segue because there were a few minutes left in class. She wanted to give the groups an opportunity to revise their representations. She told the class, "I am really happy with your hard-thinking work today! What I want you to do now is take your whiteboards back to your tables and talk with your group for five minutes. Decide if there is anything you want to change
585   about your representation and then draw your changes if you want to make any. Remember to try to use the ideas we discussed: SPACING, FORCES, MOTION, and HEAT ENERGY." She told students that when they were satisfied with their representations they could begin their homework. Because she wanted to see what students understood about what they had discussed, Ms.
590   Nichols decided to have the students respond in writing to a question related to the day's work in their science journals. She told the class, "For homework, I want you to describe something a classmate said that helped you to deepen your thinking about the features of water we discussed today. Then, explain why the idea helped you in changing your thinking. Be sure to write your answer in
595   your science journals. We will discuss your ideas tomorrow." With only a minute remaining in class, Ms. Nichols thanked the class for a good discussion, and she reminded them to return their boards to the front of the room.

Analysis of Connecting in the Case of Kendra Nichols

Ms. Nichols's intent in this lesson was for her students to learn a set of ideas related to matter and molecules regarding spacing, size, force, motion and heat (see part 3 of the Case of Kendra Nichols from earlier in this chapter, lines 283–97). Through her work in anticipating, monitoring, selecting, and sequencing, Ms. Nichols positioned herself to make connections among different solutions created by her students and with the key disciplinary ideas that were central to the lesson. In the following sections we will discuss (1) the way in which each of the key features of water targeted for discussion was made public by building on the ideas generated by students (connections to disciplinary ideas); and (2) the ways in which different representations built on and were compared to each other in order to enable students to develop understanding (connections among different solutions).

Size of molecules does not change

Ms. Nichols wanted her students to know that the type and size of the water molecule stays the same in all phases. She decided to start with the representation produced by group A since they showed molecules of the same size in all three phases. Fatima made this point in her claim, "Yes, it's all water molecules. They are all the same" (lines 378–79). Later in the lesson, group B—who had shown the molecules changing size in each phase—acknowledged that they "drew the molecules wrong" (lines 539–40) and stated, "We thought they changed size to get to the different phases by adding heat or taking away heat. But it makes sense that the molecules stay the same size" (lines 546–48).

Molecular spacing changes

Ms. Nichols wanted her students to know that molecular spacing changes from one phase to another, but molecules of water are most closely packed in the liquid phase and furthest apart in the gas phase. As four of the six groups drew representations that featured incorrect spacing for the solid and liquid phases, she decided to begin with a discussion of group A's poster (incorrect spacing) and quickly move on to group D's poster (correct spacing). By juxtaposing the two representations, students were able to see how they differed on this feature. Group A began the discussion by highlighting the fact that water molecules "are spaced differently in each phase" (line 379). Krystal from group D indicated that, while her group first drew a picture similar to the one drawn by group A (different spacing in each phase but not the correct spacing), they changed it because "we remembered that the frozen water took up more space in our experiment, so we changed the drawing and put more space between them [the molecules in frozen water]" (lines 394–95). Although Krystal's explanation was accurate, Ms. Nichols was not convinced that all students were following this and went on to engage them in a thought experiment intended to show that group A's spacing was incorrect and group D's was correct. Through this process the students arrived at the conclusion that water molecules are farther apart in solid than in liquid, closest in liquid, and furthest apart in gas. Ms. Nichols recorded this idea on the board so that students could keep track of the conclusions they had collectively reached.

Solid molecules are connected by attractive forces

Ms. Nichols wanted students to know that, in the solid state, water molecules are connected by attractive forces that hold them in a crystal structure. When motion increases, the relative effect of the forces decreases. This idea first surfaced during the discussion of group D's poster, when Ms. Nichols asked Krystal about the lines her group drew between the molecules in ice. Krystal said,

"We think that the molecules [in ice] have to push away from each other somehow because they are farther apart than the water. So whatever is pushing them apart, kind of holds them in place, too, because it is solid" (lines 453–56). This led to a discussion of force, the use of magnetic marbles as a tool for understanding attraction, the introduction of a diagram of water molecules (fig. 4.6), and the conclusion regarding force that was recorded on the board.

Motion of molecules changes

Ms. Nichols wanted students to know that the motion of molecules changes from phase to phase, with the most motion (in terms of distance and direction) happening in the gas phase and the least in the solid phase. Specifically, gas molecules spread out beyond the container of water while the molecules of ice and water are "contained"—either self-contained (solid ice) or contained by a physical boundary (the glass holding water). This idea first came up when Ms. Nichols drew students' attention to the gas molecules in group D's poster and noted that the molecules were "all over the place" (lines 505–6). She then invited Yuan from group E to share her group's ideas about motion. Yuan indicated that her group represented motion "like group D did" and went on to explain, "In the solid state, the molecules are vibrating. The liquid moves a little more, but not as much as the gas. Gases move everywhere, no specific places" (lines 509–13). Ms. Nichols then prompted students to think about the idea of force that had been introduced in the discussion of group D's poster and asked the class if the liquid and gas phases have forces between the molecules just like the solid phase. Again, students used the magnetic marbles to simulate the force when the molecules were close together. Ms. Nichols concluded this part of the discussion by saying that "just like the magnetic marbles, when water molecules are moving fast enough and are far enough away from each other, these forces are not as important. Once the molecules slow down and get closer the forces are more noticeable between the molecules (lines 529–32). She then recorded this feature on the chart she had created on the board.

Heat energy

Ms. Nichols wanted students to know that increasing heat energy results in changes of phase (e.g., from liquid to gas) and decreasing heat energy results in changes of phase (e.g., from gas to liquid). The only group to include heat in their representation was group B, although they had inaccurately represented other features of water. By having this group go last Ms. Nichols was able to introduce a feature that had not yet been discussed, provide an explanation regarding how water actually transitions from one phase to another, and give group B a chance to talk about the accuracy of their model in light of the preceding discussion. Ms. Nichols asked group B to explain the role of heat and to connect their explanation with group E's representation. This surfaced the fact that if you add heat to frozen water it melts and that if you add heat to liquid water you get gaseous water or vapor. Carlos explained that "to get the water to boil you add heat and the molecules move faster and farther and farther until they are a gas" (lines 562–64). Ms. Nichols then added the idea of heat energy to the list of features she was tracking.

Ms. Nichols's careful selecting and sequencing of student responses helped her build the story line for the discussion that started with a basic feature of water (size and type of molecules) and move on to more complicated ideas, culminating with a discussion of heat. Throughout the discussion, Ms. Nichols worked to clarify students' ideas, remind them of the experiments with water they had previously conducted, and introduce new representations (i.e., jars of marbles, magnetic marbles, an image of the structure of ice) to ensure that students understood the features of water. While Ms. Nichols clearly built on the work of her students, she did not hesitate to provide information or clarify points as needed. The use of the whiteboards made it possible for students to easily modify their representations at the end of class in light of the discussion and would provide Ms. Nichols with insight into students' current understanding of the molecular explanation for the behavior of water in different states.

Analysis of the Science Practices in the Case of Kendra Nichols

The lessons built around the Explaining the Behavior of Water task provided students with the opportunity to develop an understanding of one of the disciplinary core ideas from the Next Generation Science Standards (NGSS)—in this case, Structure and Properties of Matter (MS-PS1.A). In addition, the lessons afforded students with many opportunities to productively engage in the Science Practices described in the NGSS (Achieve, Inc. 2013). Specifically, prior to the Explaining the Behavior of Water task, Ms. Nichols's students engaged in a series of investigations (SP 3; see the list in fig. 0.1 on page 1) to explore the behavior of water (fig. 3.1). Using the patterns they noticed during the investigations, Ms. Nichols asked the students to analyze and interpret the data (SP 4) in order to develop a model (SP 2) that described the behavior of water molecules. From their model, the students then constructed an explanation (SP 6) about the behavior of water molecules (fig. 3.2; fig. 3.5; lines 136–39 in part 2 of the Case of Kendra Nichols in chapter 3).

Ms. Nichols began the discussion by having Fatima explain her group's representation (lines 362–74). In doing so, she gave that group (and others that presented) an opportunity to communicate and describe their thinking (SP 8). At the same time, Ms. Nichols pressed students to evaluate the ideas of their classmates (SP 8) (lines 425–31; lines 555–64). As the class engaged in the discussion, the students developed a consensus model (SP 2) of the behavior of water molecules in the solid, liquid, and gaseous phases and revised their own models to align with the consensus model (lines 584–86). Once the discussion came to a close, Ms. Nichols assigned homework that enabled students to describe and evaluate the ideas of their classmates (SP 8) and explain how that contributed to their own learning (lines 591–95).

Conclusion

Although anticipating ensures that teachers have thought carefully about what students might do and say, and monitoring ensures that they have paid close attention to students' thinking during the lesson, it is through selecting, sequencing, and connecting that teachers guarantee that key approaches and ways of thinking about the task are made public so that all students have the opportunity to make sense of science ideas. Although there are many ways to select, sequence, and connect responses, these decisions must be guided by what the teacher is trying to accomplish in the lesson. Hence, the goals for the lesson serve as a beacon toward which all activity is directed. As Hiebert and his colleagues (2007, p. 51) say, "Formulating clear, explicit learning goals sets the stage for everything else."

In the Case of Kendra Nichols, we see a teacher who took this idea to heart. She clearly identified her goals early in the planning process and never lost sight of them as she moved through the actual implementation of the lesson. Although we might point to things about the lesson that could be improved, Ms. Nichols's clear focus on what she wanted to accomplish led her to select, sequence, and connect the responses in such a way that the ideas with which she wanted students to grapple were in the public arena. Hatano and Inagaki (1991, p. 341) argue that "a group as a whole usually has a richer data base than any of its members for problem solving. It is likely that no individual member has acquired or has ready access to all needed pieces of information, but every piece is owned by at least one member in the group." Thus, during group discussion, all participants have the opportunity to "collect more pieces of information about the issue of the discussion and to understand the issue more deeply" (Hatano and Inagaki 1991, p. 346). Although Ms. Nichols will need to do additional work to assess what individual students took away from the discussion, the revised models and homework that she assigned is likely to provide insights that will help her in designing subsequent instruction.

Our attention in chapters 3 and 4 has been on how to use the five practices to orchestrate a productive discussion. As a result of this targeted focus, we did not explicitly address other things that contribute to the success of a lesson in which the five practices were used. In the next chapter, we focus on particular talk moves that teachers can make as they interact with students moment-by-moment in the classroom that will help to ensure that the five practices are enacted to their full potential.

TRY THIS!

- Teach the lesson you that planned at the conclusion of chapter 3. Collect data by using the monitoring sheet that you created and then indicate which solutions you will select and the order in which you will sequence the presentations.

- You may want to make an audio or video recording of the discussion so that you can reflect on the extent to which you were able to make connections among different solutions and with the mathematical ideas that were central to the lesson.

CHAPTER 5

Encouraging and Guiding Student Thinking

The five practices can help teachers to plan for and manage classroom discussions productively. However, teachers will need to develop additional skills in order to help students learn scientific concepts through engagement in disciplinary practices. We have already discussed the importance of setting appropriate performance and learning goals for students and selecting instructional tasks that provide students with the opportunity to engage in the scientific practices. In addition, teachers need to learn how to interact with students in a way that monitors, directs, and guides their thinking and that encourages the development of the scientific practices. The purpose of this chapter is to identify particular moves teachers can make as they interact with students moment by moment in the classroom that will help to ensure that the five practices are enacted to their full potential.

Why is the manner in which teachers interact with students so important that it warrants a separate chapter? *What* students learn is entwined with *how* they learn it. And the stage is set for the *how* of learning by the nature of classroom-based interactions between and among teachers and students. There are a variety of ways in which teachers might interact with students and students with one another, ranging from abrupt, short-answer question-and-answer sessions to deeper explorations of concepts and ideas. Each of these styles of interaction is associated with different opportunities for student learning.

In the first part of the chapter, we will describe three different kinds of classroom talk and how they can shape the opportunities that students have to think, reason, and make sense of scientific phenomena. Next we will provide illustrations of teachers' use of each type of talk in small-group settings as they are monitoring students' work. This is followed by illustrations of the role of each type of talk in the whole-class discussions that typically follow small-group work. We conclude by revisiting the importance of student thinking and the role of teachers in supporting that thinking.

Kinds of Classroom Talk

In the broadest sense, the goal of science education is for students to understand natural phenomena. Understanding entails finding and describing patterns, looking for and testing relationships, and making connections between cause and effect. In order to learn science with understanding, students have to *think*.

They cannot develop understandings by parroting what the teacher says or memorizing what is in textbooks. Although this sounds self-evident, thinking is often in short supply in U.S. science classrooms. Instead, what one is apt to find is students completing worksheets, memorizing facts, and reproducing definitions (Roth et al. 2006).

For some, the antidote to impoverished thinking in the science classroom is hands-on activities. This is an incomplete solution. Although these activities have the potential to elicit student thinking, even the best-designed activity does not guarantee that thinking will ensue or that understanding will develop. Despite active engagement with materials, there is no assurance that, left on their own, students will advance their understanding of the scientific world. Students often attend to irrelevant features and fail to notice important ones. Even if they do identify relevant features, they might not engage in reasoning processes fundamental to the development of understanding: finding patterns and relationships, representing information in new ways, and constructing and testing explanations.

We believe that the high cognitive demand tasks that are foregrounded in chapter 1 can go a long way toward spurring thinking. These tasks cannot be completed by following a set of prescribed "cookbook-like" directions, reading information from a text to get the correct answer, or applying a known definition or rule. Even when students are presented with high cognitive demand tasks, however, it is incumbent upon the teacher to support their thinking while they are engaged with the task and as they discuss their emerging understandings with the rest of the class. Otherwise, the teacher's actions and interactions can serve to decrease or take over the thinking, thereby reducing the cognitive demand of the task.

How can teachers help to ensure that students learn to think and reason in their classrooms? In addition to following the five practices, they need to actively surface student thinking and shape it in ways that are productive. Newton calls this *focused talk*, dialogue between teachers and students that pushes student thinking toward "what is needed to move learning forward" (2002, p. 33). Such talk is a necessary accompaniment to the tasks and practices identified in this book.

Classroom talk surrounding rich tasks like those discussed in chapter 1 is a good way to move students' engagement with ideas in a "thinking" direction. Teachers can "push" students' thinking toward deeper and more sophisticated forms by asking a well-timed question or by noting a contradiction. This kind of talk is the opposite of what is often seen in classrooms in which the ritual of "Guess what is on my mind?" is enacted, the most common form being the IRE pattern in which the teacher *initiates* a question, the student *responds* (usually in one or two words), and the teacher *evaluates* the student's response as either right or wrong (Mehan 1979). These types of exchanges do little to deepen students' comprehension of the material before them; instead, students learn to guess the known-answer to the teacher's question. Moreover, the authority for deciding if an answer is right or wrong lies solely with the teacher and not on members of the learning community (students) who are engaged in discipline-based reasoning. This leaves the student completely dependent on a "knowledgeable other" for judging the veracity of his or her answers.

Figure 5.1 (left-hand side) presents an example of IRE-style talk surrounding the genotype task that was introduced in chapter 4 (see fig. 4.1). Although it can be argued that the teacher was leading students toward the goals of the lesson (i.e., students' knowing that a genotype for a particular monogenic trait is composed of two alleles and that genotype can only be determined by knowing the genotype of both parents), the IRE dialogue illustrates *how* students were being led toward those lesson goals: in a way that invested little to no effort in building students' own capacities to reason about scientific phenomena. On the other hand, the dialogue in the right-hand portion of figure 5.1, which we are characterizing as focused talk, illustrates how a teacher could lead a discussion using the very same set of student responses and about the very same goals, but in a manner in which students are asked to think and reason for themselves instead of looking to the teacher for evaluation of whether or not they are correct.

IRE Pattern	Focused Talk
T: Who do the five children get their earlobes from? (*Initiation*) St 1: The father. (*Response*) T: No. Their earlobes look like their father's but that's not the only place the information comes from. (*Evaluation*) Does anyone else want to answer my question? (*Initiation*) St 2: They get them from both the mother and the father. (*Response*) T: Yes! (*Evaluation*) All traits are inherited from *both* the mother and the father. What is the genotype of the father? (*Initiation*) St 3: (a student from the group that produced response 2, fig. 4.1) Homozygous. We called it capital E capital E. (*Response*) T: That *might* be his genotype, but if that is all you put in your answer you wouldn't get full credit. (*Evaluation*) We cannot know for sure what the genotype of the father is. So you'd need to say that. Why can't we know for sure? (*Initiation*)	T: Let's take a closer look at your responses to the task you've been working on (see responses 1–4, fig. 4.1). The students in group 4 stated that the *bloodline comes through the father*. So they are speculating that this might account for why the kids' earlobes are the same as the father's. Would the diagram from group 2 support that? St 1: The diagram shows the kids have stuff from the mother and the father. T: Can you point to where in the diagram you see that, Carla? St 1: Inside the boxes, you can see that the children have two letters, and one is from the father and one is from the mother. So you can't say the bloodline is just from the father. T: Those letters represent alleles. What do the rest of you think? St 2: I agree with Carla. Our group (group 1) pretty much found the same thing. St 3: But you have different letters than the other group. St 2: What letter you decide to use to show the genotype of the mother and dad . . . or even the kids . . . really doesn't matter, you just have to be consistent in how you work out what the kids get. T: So getting back to group 4, how are you thinking about your response now? Have your ideas changed at all? St 4: We weren't really thinking about the different, uh—alleles. We were just noticing that the kids looked like their dad and so we thought that what they got from him was more important. T: Let's think about that idea along with what we heard already. We heard that the kids get information from both mom and dad. But these particular kids all look like dad. So how do we make sense of that?

Fig. 5.1. Examples of IRE pattern and focused talk

The Next Generation Science Standards (Achieve, Inc. 2013) aim to foster students' development into active thinkers, constructors, and evaluators of knowledge, a near impossibility under the IRE format of classroom discussion. Unfortunately, though, most teachers are comfortable with the IRE form of discussion and feel unprepared to lead discussions similar to the focused talk seen on the right-hand side of figure 5.1, the kind of discussions that would develop their students' capacity to be knowledge authorities in the science classroom. In addition, most students have not participated in discussions in which their independent thinking is valued and held accountable to disciplinary norms. They are more often used to being told the facts to memorize, the definitions to reproduce, or the steps to follow to conduct an investigation. Even when they do engage in "scientific practices," it is usually for the sake of learning/memorizing the practices rather than an authentic experience they engage in to build new knowledge. This overemphasis on learning the practice (almost to the point of obscuring the building of new knowledge) can be seen in figure 1.3a in chapter 1. As written, the task shown there heavily specifies the procedures of collecting and recording data on plant height and is silent on the science content that can be surfaced and learned.

In the next section of this chapter, we provide concrete suggestions for how teachers can support students to think more deeply about rich tasks in science. Good tasks, as noted in chapter 1, are an essential starting point. However, it is what teachers choose to attend to and cultivate *during instruction*—in students' work and in their responses to questions—that will ultimately determine whether or not students become active thinkers, constructors, and evaluators of knowledge in science.

Three kinds of focused talk will be highlighted: talk that (1) makes students' thinking visible; (2) guides students' thinking in productive directions; and (3) directs students' attention toward features of the problem space that matter. Each of these talk types will be illustrated as teachers monitor group work and as they orchestrate whole-class discussions. The group-work illustrations will be presented first and are drawn from the Case of Kendra Nichols (chapters 3 and 4). The whole-class examples that follow them are primarily drawn from the Case of Nathan Gates (chapter 2).

Use of Focused Talk While Monitoring Small-Group Work

As noted in earlier chapters, the teacher's goal when intervening in a small group is not to make sure that, by the end of the work session, all students have produced a complete and correct response. Rather, it is to support students' fledgling efforts to make sense of the task before them and to make sure that their thinking is headed in a productive direction. Later in the lesson, students will have the opportunity to compare their work with what other students have produced and to participate in a class-wide discussion that clarifies the thinking behind and the features of a good explanation for whatever task on which they are working. Also, students can be provided with time toward the end of the lesson to modify their representations based on what they have learned in the whole-class discussion, thereby finishing the lesson with a complete and accurate representation/explanation of the scientific phenomenon they are studying.

Making Student Thinking Visible

It is easy to assume that students understand when they really don't. Making thinking visible "obliges students to express or use their understanding so you can appraise it" (Newton 2002, p. 46). According to Newton (2002), making student thinking visible serves two functions: It helps teachers to know whether things are going as planned (or not), and it exposes students' current understandings so that the teacher and other students can step in and redirect them if needed. An added benefit is that the very act of explaining can help students to clarify their own thinking.

Making student thinking visible is more difficult than might be assumed. Students may not be accustomed to vocalizing how they are thinking about something and typically have a difficult time doing so at first. Also, to get students to talk about their thinking, teachers need to ask them questions as they are actively trying to make sense of a rich task. Asking questions in the midst of a low-level task or after they have just read a passage in a textbook will most likely tell a teacher how students recall information or memorize a procedure. To learn how students are making sense of scientific phenomena, teachers need to question them in the midst of their grappling with a demanding task. Another way to make student thinking visible is to ask them to create artifacts (e.g., diagrams, graphs, or models) to show their work. Sometimes the artifacts reveal inaccurate or incomplete thinking about which the teacher or other students can then follow up; artifacts can also be used to support a student's verbal explanation, as, for example, when a student points to the line on a graph as she explains how two variables relate to one another.

Some examples of questions that can help make student thinking visible: "I noticed that you decided to use the mean. Why did you use the mean?" or "How did you know to do that?" These kinds of questions force students to go beyond telling you *what* they did and explain why they did it.

Another example of questions that can surface student thinking is prediction questions. When accompanied by a press for justification, prediction questions can expose how students are thinking because they ask them to form and express a theory regarding how things work (Newton 2002). For example, a teacher might ask students to predict the time at which the first quarter moon will rise. Students might answer in one of three ways, each of which provides a different level of insight into their thinking. Some students might simply guess a time and provide no justification. Without further probing, the teacher only knows that the students are right or wrong, not whether or how they understand the underlying model of moon phases. Other students might say "around noon" and then justify their prediction by saying something like, "When we were collecting our data last week we saw that the first quarter moon was really low in the east in the early afternoon, so we think it must rise a little bit before that." This suggests that the students are predicting based on a pattern. Finally, other students might also say "around noon" but then go on to justify their prediction by drawing a diagram of where the moon, earth, and sun are with respect to one another when the moon is in its first quarter phase; they might also say that "because in the first quarter the moon is 90° from where we are when we're looking at the sun at noon, that means it's just coming up over our horizon at that time." This tells the teacher that these students understand the underlying cause/effect relationships at play here.

Knowing how students are making sense of scientific phenomena is critically important for teachers who want to help students become more independent in their thinking about and evaluating scientific knowledge. Focused talk that can bring their thinking to the surface is therefore an essential ingredient of expert pedagogy. We turn next to an illustration of how Ms. Nichols used focused talk while students were engaged with a complex task.

Illustration from the Case of Ms. Nichols—Making student thinking visible as they work in groups is supported by the teacher's goals and her anticipations regarding how students are apt to approach the task. This was the case with Ms. Nichols, who—armed with her monitoring tool—circulated among the student groups, listening in on their conversations. Whenever she noticed something that she had anticipated, she would zoom in to try to uncover how the students were thinking about it. For example, when Ms. Nichols approached group B (Kim, Kevin, Lamar, Rosa), she noticed that the students represented the role of heat energy in their model (see chapter 3, fig. 3.5, group B poster). She viewed their inclusion of heat energy as a positive sign because the role of heat was one of the features that they would need to include in their explanation. Ms. Nichols began by asking the group to explain *why* they included heat in their model. As previously noted, "why" questions are often an effective way to reveal student thinking because they ask students to go beyond simply reciting *what* they did and reveal what their underlying rationale was for doing it.

Rosa's explanation revealed that the group understood that adding heat was needed to cause certain phase changes (e.g., melting), while taking away heat was required to create the opposite phase changes (e.g., freezing). However, Rosa's response did not reveal an understanding of the connection between heat and molecular motion; this caused Ms. Nichols to suspect that their thinking might be incomplete, if not inaccurate.

Her suspicions were confirmed when she noticed that group B's diagram pictured the molecules as expanding in size while the spacing between molecules remained the same. This raised an immediate concern with Ms. Nichols, because one of the misconceptions she had anticipated was that students would think that molecules in different phases were different sizes. She asked Kevin to explain the group's thinking behind the size of the molecules. Similar to "why" questions, pressing students to articulate a rationale for a decision they've made not only helps to surface their thinking but also holds students accountable for thinking by signaling that the norm in the classroom is to strive for understanding (Newton 2002). Kevin replied that their experiment showed that adding heat caused molecules to expand. His reply was very telling; Ms. Nichols now knew that the group was not taking into account the pattern that the class had noticed in the previous investigation (see fig. 3.1, experiment 2); that is, that the volume of water increases when heat is removed and the phase changes from liquid to solid.

In Ms. Nichols' interactions with group B, we see how she posed strategic questions that uncovered how students were thinking about the underlying cause of the behavior of water. This example also illustrates how this form of focused talk serves two purposes: monitoring whether or not students are thinking accurately and, based on what the teacher discovers about how students are thinking, helping her to devise tailored assistance to revise their thinking. We will learn about that tailored assistance next.

Guiding Student Thinking in Productive Directions

An accomplished scientist's head is not filled with memorized facts. Although she will certainly know some facts, facts alone will only take her so far when confronted with a novel problem. Accomplished scientists possess knowledge that is flexible and useful in nonroutine situations; such knowledge comes from engaging in scientific practices to build understanding.

Fostering students' understanding can be thought of as scaffolding their thinking to enable them to construct meaningful and valid representations of the situations or events they are studying. Such representations often begin at a descriptive level—observing and describing regularities or patterns. To progress in scientific understanding, however, students need to advance beyond description to produce explanations that (ultimately) include cause-and-effect relationships.

One way to guide students' thinking in productive ways is to scaffold their use of scientific practices such as looking for patterns, collecting and analyzing data, and building and testing models to explain scientific phenomena. To do so, teachers need to be alert to the possibilities offered by the instructional task and to be aware of students' current understandings.

Illustration from the Case of Ms. Nichols—Exactly how the teacher guides student thinking is shaped by what she uncovers as she monitors student thinking. For example, in the previous section, we left Ms. Nichols as she discovered that Kevin's group was not taking into account the pattern that the class had identified in the previous investigation (see fig. 3.1, experiment 2); that is, that the volume of water increases when heat is taken away and the phase changes from liquid to solid. Ms. Nichols saw this as an opportunity to intervene. However, rather than telling the students the correct way to think, Ms. Nichols asked another group member to summarize the findings from experiment 2. Lamar said that the vial with frozen water had a bigger volume than the ones that were just sitting in the room or in the refrigerator. Ms. Nichols then asked Kevin's group to consider if their picture was consistent with this finding and if not, how they might adjust it. She left the group to ponder the question.

This intervention was critical in guiding student thinking into a more productive channel. Ms. Nichols's goal was *not* to correct student thinking or to efficiently guide their work into a complete and accurate representation of the behavior of water molecules, at least not at this point. Students would have the opportunity to further refine and (if necessary) correct their knowledge during the whole-class discussion. Ms. Nichols' goal was instead to get students to consider additional pieces of information—more specifically, discrepant information—that might coax their sense making into a more fruitful space where they might straighten out their misconceptions on their own.

In a task like this one, the work of the scientist is to provide an explanation that persuasively accounts for the patterns that were revealed in prior experiments. Accordingly, we see Ms. Nichols pushing the students to "test" their early ideas by seeing if they can explain the data patterns previously noticed (Stewart, Cartier, and Passmore 2005; Cartier et al. 2005). Ms. Nichols' suggestion to incorporate the results of experiment 2 into their thinking scaffolds students into the scientific practice of testing their model against the available empirical evidence. It requires hard thinking—but the payoff will be a fuller, more accurate understanding of the behavior of water molecules.

Directing Students' Attention to What Matters

Natural phenomena are complex. They can be described, represented, and analyzed in many different ways, some of which are more helpful than others. When students first engage in rich, problem-based tasks, they do not necessarily possess the orienting frameworks that would lead them to notice and represent the parts of the situation that are consequential for the problem they are trying to solve. Instead they focus on irrelevant features that can lead them astray. For example, while making observations about how the position of a shadow changed as time passed, Newton (2002) found that young students focused on the moving clouds, integrating them into their explanation of why the shadows moved instead of noticing and using the changing position of the sun over time. In this case, focused talk was needed to direct students' attention to three variables: time, the positioning of the shadows, and the position of the sun.

Illustration from the Case of Ms. Nichols—Sometimes students have difficulty abstracting *any* features of a situation that could aid scientific thought. Such was the case with group F in Ms. Nichols's class. When asked to provide a model of what they saw through the Magic Science Glasses, these students drew a picture of an ice cube, a puddle of water, and steam coming from a tea kettle; these pictures reveal nothing about what was happening at the molecular level. These students needed immediate help refocusing their thinking on features of the problem space that would allow them to make progress on their task: to depict a model of the behavior of water molecules that could explain the patterns that they observed in their earlier experiments.

Ms. Nichols began by revisiting the data table summarizing what they noticed in their earlier investigations. In so doing, she was trying to refocus students' attention away from everyday observations (a boiling tea kettle) and toward the level of molecular behavior that the Magic Glasses were meant to make visible. In particular, she wanted to draw their attention to the important features of the experiments that she knew would be key if they were going to be able to construct an explanation: the spacing between molecules, attractive forces, the motion of the molecules, and the role of heat energy. The dialogue that follows was first seen in chapter 3 (part 2 of the Case of Kendra Nichols, lines 181–205). In the version that follows, the features of water that Ms. Nichols was attempting to draw attention to have been identified in bold type.

Ms. Nichols:	Okay, so let's look back at our experiments. What pattern did we notice in experiment 1?
Stefan:	(*Reads from chart displayed at the front of the room*) "Solid water retains a definite shape and can't be compressed using a syringe. Liquid water flows to the bottom of the container and is not measurably compressed in the syringe. Gaseous water escapes (disperses) out of the container and can be compressed when put into a syringe (when we push down on the plunger)."
Ms. Nichols:	Right, so what does that mean? Can someone explain that in your own words?
Lei:	Solid water is hard and can't be pressed when squeezed **(spacing)** and liquid water is moveable and the gas moves a lot **(motion).**
Ms. Nichols:	Okay, Stefan, does that make sense from what you read?

Stefan:	Liquid water moves around (**motion**) and solid is hard and can't be squeezed (**spacing**). Yes.
Ms. Nichols:	All right. And what about in experiment 2? Marcellus, tell us in your own words what pattern we noticed in experiment 2.
Marcellus:	The volume of water at the freezing temperature changed (**role of heat energy**), got bigger (**spacing**). The warmer vials of water all had the same volumes.
Ms. Nichols:	Okay, Yvette. Do you agree with Marcellus's summary of the pattern we noticed from experiment 2?
Yvette:	Yes.
Ms. Nichols:	Yes? Okay, Yvette, I want you to summarize experiment 3 for your group just like we did for 1 and 2. Remember, we are looking for patterns. Then, drawing on these patterns, I want the group to create a representation that shows what water looks like when we can see its particles with our magic glasses on . . . Okay? (**This experiment raises the need for an explanation regarding attractive forces.**)

By reviewing the findings from the previous day's experiments, Ms. Nichols meant to draw students' attention to the features of the situation that matter in the construction of an explanation regarding the behavior of water molecules. Perhaps the least obvious feature was the role of attractive forces, so before leaving the group, Ms. Nichols suggested that the students use the magnetic marbles in their baskets as a tangible representation to visualize the interactions between water molecules.

Later in the same lesson, as students presented their diagrams to the whole class, Ms. Nichols became even more explicit in clearly marking the features of a good explanation for the behavior of water molecules. Because she knew that none of the groups had created a representation depicting all the necessary features, and she wanted to scaffold students' ability to correctly modify their representations at the end of the class period, she created a table of important features for the whole class to see and use. As students' presentations touched upon a particular feature, Ms. Nichols added that feature to the table (see fig. 4.5). This table was then used as a touchstone for students near the end of the lesson when they were directed to go back to their diagrams and to "fix them" so as to be a complete and accurate representation of the behavior of water molecules.

Use of Focused Talk during Whole-Class Discussions

Unlike the use of focused talk during small-group work, the point during whole-class discussions is to make sure that a complete and correct product comes out of the discussion. This lends a slightly different quality to how teachers use focused talk in these situations. Most noticeably, there is more "shaping" of student thinking and a slightly more pronounced effort to guide the class as a whole toward achieving a satisfying and accurate conclusion to the lesson.

Making Student Thinking Visible

Classrooms in which student thinking is regularly surfaced and made public during whole-class discussions provide a supportive environment in which students can learn valuable skills. The student who is explaining has the opportunity to practice communicating her ideas to others, an important scientific practice as outlined in the Next Generation Science Standards (NGSS). The nonspeakers benefit, too. When students know that they may be asked at any time to add to the discussion, they learn to listen to other students' explanations and to actively critique and compare them to their own developing ideas.

Student thinking is important in whole-class discussions for another reason. As noted earlier, it is critical that a complete and correct product emerges out of the whole-class discussion that occurs near the end of the lesson. Yet, in the Five Practices model, teachers do not supply the complete and accurate product but rather use student thinking as fodder for moving the entire class toward a complete and accurate understanding of the scientific learning goal of the lesson. Doing so requires some strategizing regarding who is selected to present and when and what kinds of questions to ask to put students in "the right space" to begin to grapple with important issues.

Illustration from the Case of Mr. Gates—In chapter 2, Mr. Gates wanted students to understand that range and shape are two important features for describing typical Fast Plant growth. Thus, he chose group 7 and pursued a line of questioning that exposed student ideas that he could work with (see fig. 2.3).

Group 7's graph depicted the growth of all six of their Fast Plants. A spokesperson for group 7 (Ryanne) explained that the group measured the height of each plant and found that from day 13 to 21 the plants "grew a lot." So, she explained, they chose to represent their data in a line graph that depicted the growth of all six of their Fast Plants.

In the excerpt from the Case of Nathan Gates that follows (chapter 2, lines 95–105), the questions posed by Mr. Gates that are now in bold served to surface how students were thinking about the task.

Mr. Gates then posed a question to the class: "**What are some things you notice** about the representation group 7 has created?" Several students shared their ideas:

Juan:	You can easily see the day of measurement and the height of the plants.
Mr. Gates:	Okay, Juan, **where do you see that?**
Juan:	The graph has axes that are labeled and there is a key so we can tell which plant is which.
Mr. Gates:	Okay, so the x and y axes allow us to understand what data is represented. Class, **do we agree with that?**
Trina:	I do, you can also see the height of all the plants on any day they were measured.

The questions Mr. Gates asked were very open, beginning with: "What are some things that you noticed?" His rejoinder to students' responses served to further clarify or elaborate student thinking ("Where do you see that?") or to invite other students into the conversation ("Do we agree with that?").

However, as can be seen in the following excerpt from that case (lines 106–11), Mr. Gates also worked to *shape* student thinking toward one of the goals of the lesson: recognizing that shape is an important attribute to include when describing typical Fast Plant growth.

Mr. Gates:	Okay, so **what does this graph tell you about the plants' growth?**
Trina:	The plants get taller over time.
Mr. Gates:	Okay, the plants get taller over time. **What else?**
David:	Some plants are growing faster and taller than others.
Tessa:	The plants start out growing slowly, then they really grow a lot, and then they sort of don't grow much.

In this set of questions, Mr. Gates continues to surface and probe student thinking, but he also capitalizes on Tessa's remark by asking the class if they could "see" what Tessa described in the graphs. When another student noted that all of the graphs had the same basic S shape, Mr. Gates asked the class to validate if this S shape was true for other graphs that different groups had produced. The students agreed, and the class was on its way toward understanding the importance of growth curves in describing plants or populations of organisms. The point here, however, is that the idea of *typical growth* had not been specifically raised by the first group. By asking open-ended questions and "listening" for certain pieces of knowledge on which to build, Mr. Gates was able to get students to consider an S-shaped growth curve as a way to describe typical growth.

Guiding Student Thinking in Productive Directions

Similar to how students' thinking can be guided while they are working on tasks in small groups, students' thinking can be guided during the whole-class format. One way in which we've seen this happen is through an extension activity that builds on students' newly achieved knowledge that is of a descriptive nature. Once students have adequately described a set of patterns inherent in empirical data, for example, the teacher might push them to develop an explanation for why the pattern exists.

Illustrations from the Cases of Mr. Gates and Ms. Nichols—In the case of Mr. Gates, the extension activity that pushed student thinking beyond description to the development of an explanation was given as a homework assignment. Students were told to answer the question, "How can you account for the S-shaped growth curve?" Mr. Gates made clear that he wanted to see a written answer and a rationale for their conclusions. His plan was to use the question to launch the discussion in the next class. Ultimately, he was working toward their developing an understanding that the growth patterns can be explained by an analysis of where the plant is spending its energy resources at various stages of its life cycle. This episode illustrates shaping student thinking by building on what students have just achieved and pushing it one step further. During the next day's lesson, Mr. Gates will have the opportunity to assess what students have done and to further shape it in the direction of his learning goal.

Sometimes student responses do not provide sufficient fodder for the teacher to call upon in order to move the class in a particular direction. If the goal is to shape student thinking (i.e., *not* to provide the information directly), using a thought experiment can be useful. When Ms. Nichols realized that many students were struggling with molecule spacing in liquid vs. solid water (i.e., they illustrated the ice molecules as being closer together than the liquid water molecules—see fig. 4.2, groups A, B, C, and F), she rechanneled their thinking in a more accurate direction by engaging them in a thought experiment. She asked them to imagine what would happen when more marbles were squeezed into a glass jar. (There would be more marbles for the same amount of space.) In so doing, she was able to provide a new concrete representation that was well suited to helping students come to grips with an essential discrepancy that they would have to resolve: If there were indeed more ice molecules squeezed into the same amount of space as there were liquid water molecules (as shown in their posters), then ice in a cup of water would sink to the bottom (but it doesn't). Confronting this discrepancy, in turn, laid the groundwork for the need for an explanation that would account for the fact that ice floats in water. The class eventually co-constructed the correct explanation, that the crystalline structure of ice (see fig. 4.6) holds ice molecules in place, in a position further apart from one another than they are in liquid water.

Directing Students' Attention to What Matters

Directing students' attention to salient features of the problem space is also important in whole-class discussions. We have already seen how Ms. Nichols did so during the whole-class discussion of the behavior of water molecules when she created a chart (fig. 4.5) that marked the important features that she wanted students to attend to when they "corrected" their representations.

Illustration from the Case of Mr. Gates—The need to steer students' attention toward features of the problem space that mattered for characterizing the typical growth pattern for Fast Plants arose in relationship to group 1's representation of plant growth (see fig. 2.4). Mr. Gates chose to draw attention to this poster in order to drive home the point that *what* scientists select to represent and *how* they represent it influences their ability to communicate clearly with others, an important scientific practice. When asked to explain his group's representation, Peter explained that they too represented the height of each plant, but, unlike others, they didn't make a graph. He explained that each pot represented a plant and that each of the stems in the pot represented the height the plant was on a certain day.

When Mr. Gates asked the class what they noticed about group 1's graph, the following exchange (lines 137–61 from his case in chapter 2) unfolded:

Mitch:	Um, well, it's hard to see the measurements and hard to compare the plants.
Mr. Gates:	Okay, Mitch says that it is hard to see the measurements and difficult to compare the plants with this representation. Does anyone else agree?
Students:	Yes!
Mr. Gates:	Okay, Marie, you said yes, what makes this representation difficult for you to understand?
Marie:	I think the pots and the leaves and the plants make it really confusing. The days aren't in order and I don't think the stems in the pots are equal to the real pots. So it is hard to compare across pots. It is really unorganized.
Mitch:	I agree. Ryanne's group's graph is easier to read. We can see stuff easily. You can see how tall each plant is each day they measured and you can compare the heights of the plants on the same day.
Mr. Gates:	Okay, Mitch says we can see stuff easily in the graph. What does that tell us about representing scientific data?
Peter:	I guess graphing our data would have made it easier for everyone else to see. And, they would have an easier time making comparisons from one day to the next. But it's pretty the way we did it! And you can see other stuff like when we got flowers.
Mr. Gates:	Okay, Peter, great points. You were showing more information than just height because you drew the flowers in, too. And, it might be very easy for you to see and understand your own data in a representation like this, but it can be difficult for others to interpret. This very idea is why in science it is important to use standard representations, like a line graph or even a bar graph, to represent data. It allows us to easily see and interpret the data, especially when it is data that we didn't collect.

This dialogue illustrates how Mr. Gates was able to gently steer the class in general and Peter's group in particular toward the realization that, when illustrating scientific data, it is important to abstract and mark the features that will help others to easily compare those elements that are critical to making scientific generalizations. Although the dialogue ends with an endorsement of graphs as a valuable way to interpret data, the students were also wrestling with *what* to pay attention to. Peter's group was attending to the artistic features of their representation as well as when the flowers appeared (in addition to their measured height on specific days). Their efforts to include these features (coupled with their decision to organize data on plant height and measurement date by individual plants in isolation from one another) led to a less-than-optimal representation for answering the question: "What is the typical growth pattern for Fast Plants?" Although less pointed than Ms.

Nichols' marking of features in a publicly displayed table during the lesson on the behavior of water molecules, Mr. Gates was similarly helping students to learn to attend to the important elements for the task at hand.

Conclusion

This chapter has presented ways in which teachers can support and shape student thinking during the kinds of lessons in which the five practices are used. Such lessons feature rich, open-ended tasks and are likely to spawn divergent responses among students. As such, lessons in which the Five Practices are used demand more active guidance from the teacher than do "stand-and-deliver" lectures in which the students' obligation is to do little more than follow and reproduce the facts as expounded upon by the teacher.

In lessons in which the five practices are used, teachers are called upon to actively manage student *thinking*. In order to do so, we argue, they have to first expose it by establishing the norm that discussing one's thinking—and listening actively to other students' thinking—is what is done in the science classroom. When teachers do not surface, listen to, and monitor student thinking, they have no access to a whole layer of activity that is going on in their classrooms. Without access to how students are thinking about the task, teachers have little hope of guiding it in productive directions. In this chapter, we've also discussed and provided illustrations of several ways in which teachers can steer student thinking in worthwhile directions, including scaffolding students' use of scientific practices and marking important features of the problem space.

A tension that is inherent in student-centered pedagogical approaches is the degree to which teachers listen to student thinking while still actively shaping its direction. Too much reliance on only what students are thinking can lead to lessons in which "anything goes." Too much reliance on teacher shaping of student thinking can quickly transform a lesson into a lecture on "how to do the task." Another tension is managing time. Student-centered lessons typically take more time to cover the same amount of learning territory than do teacher lectures of the same material. Teachers need to learn how to manage trade-offs between the depth of learning that likely transpires from student-centered lessons and the benefits of covering material quickly and efficiently. In the next chapter, we take up this challenge by introducing the idea of teachers as active agents in designing instruction.

Positioning Five Practices Discussions within Instructional Design

Throughout this book, we have argued that teachers should provide opportunities for students to engage in science practices (e.g., analyzing and interpreting data, constructing explanations) while learning key disciplinary concepts. We have illustrated how a lesson that is centered around discussion, and for which the teacher utilizes the Five Practices model and strategic talk moves to plan and support instruction, can enable students to engage in various science practices while learning core ideas (see chapters 4 and 5). As noted in chapter 5, this type of classroom activity, which the Five Practices model is intended to promote, is in fact uncommon in U.S. science classrooms (Roth et al. 2006). Further, we acknowledge that familiarity with the framework and a desire to enact Five Practices discussions is unlikely to be sufficient to enable teachers to create and support robust learning opportunities for students. (By "Five Practices discussions," we mean whole-class discussions within which students construct consensus understanding of underlying learning goals by sharing artifacts of their work on a challenging task and in which the teacher carefully orchestrates that sharing by using the Five Practices model to anticipate, monitor, select, sequence, and connect elements of students' work.) To sustain truly effective learning opportunities, teachers must learn to draw upon the framework, and other resources such as the Learning Cycle framework, as they engage in instructional design—a daily process that involves purposeful use of curriculum resources; consideration of students' prior knowledge, experiences, and interests; and careful selection (or design) and sequencing of learning opportunities for students.

We begin this chapter by describing the Learning Cycle and its relationship to inquiry practices. Next, we describe how teachers can draw upon the Learning Cycle to create coherent *sequences* of lessons that include opportunities for students to engage in Five Practices discussions. Finally, we present a view of teachers as instructional designers and examine how teachers' consideration of resources and constraints might inform their choices about positioning Five Practices discussions within particular Learning Cycles.

The Learning Cycle

In the late 1950s, scientists and science educators endeavored to reform the K–12 science curriculum with the goal of preparing more U.S. citizens for careers in science and engineering (DeBoer 1991). Like today, the heart of this reform effort was the commitment to enabling students to learn science through participation in inquiry (Schwab 1958). Educators recognized that authentic science inquiry was often not possible (or even desirable) for K–12 students, mostly for pragmatic reasons. However, engagement in key elements of inquiry was a reasonable goal. The Learning Cycle, first described by Atkin and Karplus (1962), provided teachers and curriculum designers with a framework for designing (or selecting) and sequencing tasks such that students would have opportunities to *Explore* phenomena, *Invent* cause-and-effect mechanisms, and *Discover* canonical explanations. In subsequent years, researchers demonstrated that lessons organized in a sequence consistent with the Learning Cycle provided students with opportunities to engage in aspects of scientific inquiry while learning core concepts (see, for example, Bybee et al. 2006; Lawson 2001; Lawson, Abraham, and Renner 1989).

Curriculum designers have proposed numerous variations to the original Learning Cycle framework. These modified frameworks include the 5E framework of *Engage, Explore, Explain, Elaborate,* and *Evaluate* (Bybee 2002); the 7E framework of *Elicit, Engage, Explore, Explain, Elaborate, Evaluate,* and *Extend* (Eisenkraft 2003); and the 4E framework of *Engage, Explore, Explain,* and *Extend* (Marshall 2007; Marshall, Horton, and Smart 2009). Although these various iterations have unique features, all Learning Cycle frameworks share the common core of *Engagement* with an initial question, *Exploration* of phenomena, and development of an *Explanation* that can account for phenomenological patterns. Most also include an opportunity for students to *Apply* new knowledge in a novel context. For the remainder of this chapter, we will use a Learning Cycle framework that includes the stages of *Engage, Explore, Explain,* and *Apply*.

Learning Cycle stage	Key inquiry experiences for learners
ENGAGE	Pose or become familiar with a question Reflect upon and articulate relevant prior knowledge or experiences
EXPLORE	Gather firsthand data and/or analyze existing data Transform and/or represent data to make patterns visible Describe patterns/trends in data
EXPLAIN	Develop and critique explanations that account for data patterns
APPLY	Use new explanatory knowledge to account for or predict new phenomena/patterns

Fig. 6.1. The Learning Cycle in science

Each stage of the Learning Cycle can be characterized by the main activities in which learners engage (fig. 6.1). During the *Engage* stage, learners either pose a question or become familiar with one that is provided for them. Also during this stage, learners have the opportunity to reflect upon and articulate their initial ideas about the topic at hand. During the *Explore* stage, learners develop descriptive knowledge about phenomena and patterns related to phenomena. This occurs either through firsthand exploration (i.e., data collection) or through analysis of existing data sets. Learners often have opportunities to use mathematical processes to transform and represent data during this stage as well. After learners describe patterns in the *Explore* stage, they develop (or are provided with) explanations for those patterns in the *Explain* stage of the Learning Cycle. Finally, the Learning Cycle often ends with an opportunity for learners to *Apply* their new knowledge by making predictions about expected patterns or explaining new ones.

As noted earlier, the Learning Cycle is an approximation of the complex process of engaging in scientific inquiry. It includes the essential activities of posing (or becoming knowledgeable about) a question, articulating relevant prior knowledge, exploring phenomena, identifying patterns, and developing explanations for those patterns. Moreover, it is easy to see how a series of lessons consistent with the Learning Cycle framework can provide students with opportunities to engage in the eight Science Practices described in the Next Generation Science Standards (NGSS) (Achieve, Inc. 2013). This relationship between practices and stages is summarized in figure 6.2.

NGSS Science Practice	Learning Cycle stage			
	ENGAGE	EXPLORE	EXPLAIN	APPLY
(SP 1) Asking questions	√			√
(SP 2) Developing and using models			√	√
(SP 3) Planning and carrying out investigations		√		
(SP 4) Analyzing and interpreting data		√	√	
(SP 5) Using mathematics and computational thinking		√		
(SP 6) Constructing explanations			√	√
(SP 7) Engaging in argument from evidence			√	√
(SP 8) Obtaining, evaluating, and communicating information		√	√	√

Fig. 6.2. Opportunities for students to engage in the Next Generation Science Standards practices (Achieve, Inc. 2013) throughout the stages of the Learning Cycle

Five Practices Discussions and the Learning Cycle

When might a Five Practices discussion occur within the context of a Learning Cycle? This depends in part upon the nature of the tasks that the teacher has chosen or designed. Recall from chapter 1 that three particular types of tasks are conducive to both high cognitive demand and engagement in Five Practices discussions: *experimentation*; *data representation, analysis and interpretation*; and *explanation* (see figs. 1.3 through 1.5). Each type of task could be expected to correspond to a particular stage (or stages) of the Learning Cycle as shown in figure 6.3. Thus, for example, if a teacher chose a demanding data representation, analysis, and interpretation task such as the Fast Plants task that Mr. Gates provided for his students (see chapter 2), the Five Practices discussion would likely take place during the *Explore* stage of the Learning Cycle. This is because data representation and analysis tasks enable students to identify patterns, and pattern recognition is the chief purpose of the *Explore* stage of the Learning Cycle. Alternatively, a Five Practices discussion centered around an explanation task (such as the one that Ms. Nichols [chapters 3 and 4] chose during the Explaining the Behavior of Water lesson) would most likely occur during the *Explain* stage of the Learning Cycle, as the purpose of such a task is to enable students to account for patterns they identified in the preceding *Explore* lessons. Generally, Five Practices discussions are productive when they occur during the *Explore, Explain,* or *Apply* stages of a Learning Cycle, after students have had an opportunity to work on a challenging task and propose a variety of strategies or models, and are therefore ready to progress collectively toward the target disciplinary learning goals.

Task type	Learning Cycle stage			
	ENGAGE	EXPLORE	EXPLAIN	APPLY
Experimentation		√		
Data Representation, Analysis, and Interpretation		√	√	√
Explanation			√	√

Fig. 6.3. Alignment between Learning Cycle stages and task types described in chapter 1

To further illustrate how a Five Practices discussion might be embedded into a science Learning Cycle, in the sections that follow, we discuss two examples in detail. The first is situated in a tenth-grade chemistry class and the second in a ninth-grade biology class.

Example 1: Studying properties of water in Mrs. Duncan's tenth-grade chemistry class

In this Learning Cycle (fig. 6.4), which spans four instructional days, the teacher, Mrs. Duncan, utilizes a variety of activity structures (see chapter 1 for a discussion of activity structures) to support learning. These include turn & talk (pair/share), small-group laboratory, interactive lecture, whole-class instruction, collaborative group work, and Five Practices discussion.

	Learning Cycle Stage	Activity Structure
Day 1	ENGAGE	**Turn & Talk (Pairs)** Mrs. Duncan poses the scenario: *My brother has a rustic cabin in the Laurel Highlands. Last year he went up there to do some snowshoeing in late January and he found that one of his bathroom pipes had cracked and was now leaking water all over the floor. He said it was because it had gotten so cold the week before and that was what made the water pipe burst. Talk to your neighbor about this. Does it make sense that freezing water would break a pipe? Have you seen anything like this before?* **Whole Class** After a few minutes, Mrs. Duncan conducts a brief whole class discussion to elicit and summarize students' initial ideas.
	EXPLORE	**Small Group Laboratory** Mrs. Duncan shares a partially completed data table with the students: _table below_ She then distributes the vials to each group and the students measure volume and mass for the frozen samples ("time 2") and enter their data into the class chart. **Whole Class** Mrs. Duncan asks for volunteers to describe patterns that they notice. She summarizes key patterns on the board: When liquid water is frozen (becomes solid), its mass does not change, but its volume increases by ~10%.
Day 2	EXPLAIN	**Interactive Lecture** Mrs. Duncan distributes magnetic marbles to each student. Using Powerpoint slides, she presents visual representations of the crystalline structure of ice and describes how hydrogen bonds hold the water molecules in this stable structure. She invites the students to use the marbles to "feel" attractive and repulsive forces that are analogous to those contributing to the overall structure of ice.
Day 3	APPLY	**Collaborative Groups** Mrs. Duncan places students into teams and provides them with a scenario: *The law firm of Dewey & Sewem have approached your engineering team to ask if you will provide expert testimony in their client's lawsuit against the Rockhard Concrete Company. Their client, Mr. Murphy, is suing Rockhard because his cement driveway has cracks in it. He thinks that Rockhard used a poor product when they installed his driveway. But Rockhard says that all cement cracks – this is a normal process in a place like Pittsburgh that experiences many freeze/thaw cycles each season. Your job is to find out about the process of laying cement and use your knowledge of the behavior of water to make an argument that either supports Mr. Murphy or Rockhard.*

Vial	VOLUME Time 1 *Room Temp.*	MASS* Time 1 *Room Temp.*	VOLUME Time 2 *Frozen*	MASS* Time 2 *Frozen*
1	25.0 ml	38.2 g		
2	25.0 ml	37.9 g		
3	25.0 ml	38.4 g		
4	25.0 ml	37.8 g		

* Mass of vial and water.

Fig. 6.4. Learning Cycle Example 1: Studying properties of water in Mrs. Duncan's tenth-grade chemistry class

	Learning Cycle Stage	Activity Structure
Day 3	APPLY	In small groups, students research cement curing processes and prepare an argument, supported by diagrams and/or pictures, to share with the whole class. While the students are in groups, the teacher monitors their work and gathers information to prepare for discussion on Day 4.
Day 4	APPLY	**"Five Practices" Discussion** *[Explanation Task]* Mrs. Duncan selects various student groups to present their arguments to the class, sequencing them carefully so that all key ideas emerge during the discussion. Students support their claims with evidence and offer critiques for others' claims. **Individual Writing Task** Mrs. Duncan poses the frozen water pipe scenario again and asks each student to write an explanation for how the pipes might have burst.

Fig. 6.4. Continued

On day 1, Mrs. Duncan begins by eliciting students' prior knowledge and experience in a brief *Engage* task, asking them to turn and talk to a peer and then to share their thoughts about a scenario in which freezing water causes pipes to burst. Next, in the *Explore* stage, the students work in their laboratory groups to gather data about the mass and volume of water samples that have been frozen overnight. Mrs. Duncan then provides the class with the initial mass and volume data (prior to freezing), and the students enter their new data into the class chart. This first day of instruction concludes with a whole-class discussion about what patterns students notice in the data, emphasizing how the mass of the frozen samples remained constant but the volume increased by about 10 percent.

On the second day of instruction, Mrs. Duncan provides each student with a pair of magnetic marbles and presents a lecture in which she describes the properties of water molecules. She uses various representations (including the marbles) to help students understand how hydrogen bonds hold water in a rigid crystalline structure, keeping the molecules a bit farther apart than they are in the liquid form. This interactive lecture makes up the *Explain* stage of the Learning Cycle.

On day 3, Mrs. Duncan begins the *Apply* stage by presenting the students with an opportunity to use their knowledge about the properties of liquid and solid water to explain a novel scenario. Working in their collaborative groups, students complete some independent research related to cement curing and develop representations to share with the class in order to support their claims about whether cracks in cement are due to poor-quality materials or to the normal behavior of water. Finally, on the fourth day of the Learning Cycle, Mrs. Duncan orchestrates a Five Practices discussion, selecting particular groups to present their arguments to the class and highlighting the key ideas about the behavior of water and how this information can be used to explain new phenomena related to cement. The Learning Cycle closes when Mrs. Duncan asks each student to return to the *Engage* scenario and to use their understanding of water molecules to explain why frozen water pipes might burst.

In this example of a Learning Cycle, students have the opportunity to develop disciplinary knowledge. Specifically, they develop **knowledge of the phenomenon** (the volume of a sample of liquid water will increase by about 10 percent when frozen) and **knowledge of the explanation** (the molecular structure of ice and the role of hydrogen bonds in forming this structure). They also have opportunities to engage in many of the science practices (SP) described in the NGSS. For example, within the *Engage* activity, students are presented with a *question* ("Why do water pipes sometimes burst when they freeze?"), related to SP 1. During the *Explore* stage of the Learning Cycle, students *carry out an investigation* (gather mass and volume data), as in SP 3; and they *analyze and interpret data* (identify patterns related to the mass and volume of liquid and solid water), as in SP 4. In the *Explain* stage, Mrs. Duncan presents the students with a model (the molecular structure of water) that accounts for the patterns they observed previously. Then, in the *Apply* stage, the students have opportunities to *use this model* (SP 2) to *construct an explanation* (SP 6) for a new phenomenon (cracking of cement). By preparing for and engaging in the Five Practices discussion, the students have an opportunity to *obtain, evaluate, and communicate information* (SP 8).

Example 2: Studying plant growth curves in Ms. Goldman's ninth-grade biology class

This second example of a Learning Cycle (fig. 6.5) involves the study of Fast Plant growth and spans eighteen days. However, during the majority of this time (days 3–15) students spend only a few minutes each day gathering Fast Plant data, and the remainder of the class time is devoted to other topics.

Throughout the Learning Cycle, students work extensively in collaborative groups. They begin in the *Engage* stage by examining samples of Fast Plants and noting various properties (e.g., stem length, leaf size). Drawing upon their observations, Ms. Goldman tells the class that they might measure these properties as a way of learning about Fast Plant growth. She then provides a protocol for measuring stem length and asks each group to design a data table to keep track of their measurements. Over the next several days (during the *Explore* stage) the students gather and record data on the growth of their plants. Then, on day 16, Ms. Goldman reminds the students that their overall goal is to be able to describe the growth of a typical Fast Plant. Similar to what Mr. Gates asked of his students (chapter 2), she asks them to work in their groups to develop a representation of their data that will enable them to answer this question. On the next day she orchestrates a Five Practices discussion in which students share and critique one another's representations and develop descriptive knowledge about growth patterns (e.g., each group of plants can be described by a range of stem length values, and each individual growth curve is roughly S-shaped). This Learning Cycle ends with the *Explain* stage on day 18, when Ms. Goldman presents a lecture to the students describing how the plant uses energy during each stage of its growth and how this use of resources accounts for the S-shaped growth curves.

	Learning Cycle Stage	Activity Structure
Day 1	ENGAGE	**Collaborative Groups** Ms. Goldman two Fastplant samples at each group work station. She selects plants that have observable differences (e.g. stem length, leaf size or abundance, etc.). She prompts students to work in their groups to describe the plants by noticing all the ways in which they are similar and different. Students list their observations in their group whiteboards. **Whole Class** After several minutes, Ms. Goldman conducts a whole class discussion about what students noticed. She creates a list of features on the board. Then she asks, *How do you think we could measure the growth of a Fastplant?* She notes that the class is going to begin a study of Fastplant growth and that it would be possible to measure growth in a variety of ways. She connects to their list of features to provide examples (e.g. *We might track the number of leaves produced by the plant or measure the change in its mass over time.*)
	EXPLORE	Then she tells the students that the way they will measure growth will be to measure stem length over time and she provides them with a protocol similar to that shown in Fig. 1.3(a). She emphasizes that over the next few weeks each group will gather growth data on their plants and then the class will get together to try to answer the question, *How does a typical Fastplant grow?* **Collaborative Groups** Each group designs their own data table and, when it is approved by Ms. Goldman, gathers stem length data for a sample of 6 Fastplants provided to them.
Days 3-15	EXPLORE	**Collaborative Groups** Students gather stem length data every 2-3 days for the first few minutes of class. (The remainder of the class period is spent on other instructional activities.)
Day 16	EXPLORE	**Collaborative Groups** Ms. Goldman reminds the students that they want to answer the question, *How does a typical Fastplant grow?* She tells each group to use their data to answer the question and to prepare a representation of their growth data to share with the class. She emphasizes that the students can mathematically transform their data if they want to and they can graph all or part of the data set as long as they can provide a rationale for their choices. The students work in their groups to determine how to represent "typical" growth using their data. While they do this, Ms. Goldman monitors their work to prepare for the discussion on Day 17.
Day 17	EXPLORE	**"Five Practices" Discussion** *[Data Representation & Analysis Task]* Ms. Goldman selects various student groups to present their arguments to the class, sequencing them carefully so that all key ideas emerge during the discussion. Students describe and support their representational choices and challenge others to do the same. Ms. Goldman uses marking strategies to summarize the key patterns and ideas that emerge from the discussion (e.g. the shape of the growth curve is roughly s-shaped; there is a range of stem length in each sample of 6 plants; using mean to describe typical is sometimes problematic because this measure of central tendency is sensitive to outliers, etc.).

Fig. 6.5. Learning Cycle Example 2: Studying plant growth curves in Ms. Goldman's ninth-grade biology class

	Learning Cycle Stage	Activity Structure
Day 18	EXPLAIN	**Interactive Lecture** Ms. Goldman uses text and visual representations to present information about how the s-shaped growth curve of a "typical" Fastplant is related to its energy use during various stages of its life cycle.

Fig. 6.5. Continued

In this second Learning Cycle example, we again see that students have opportunities to participate in multiple science practices. In particular, the *Explore* task enables them to *carry out an investigation* (SP 3); *analyze and interpret data* (SP 4); *use mathematical and computational thinking* (SP 5); and *evaluate and communicate information* (SP 8). The *Explain* portion of the Learning Cycle provides an opportunity for the students to evaluate the explanation that the teacher has constructed.

As we look across these Learning Cycle examples, it is important to emphasize that a Five Practices discussion can occur during any stage of the Learning Cycle, but most often fits within the *Explore, Explain*, or *Apply* stages. After students have had an opportunity to engage in a demanding task that affords multiple approaches or solutions and to create artifacts that represent their thinking, there will be a rich array of student thinking to draw upon in a Five Practices discussion. The focus of such a discussion might be to develop a consensus protocol for data collection (see, for example, the experimentation tasks in fig. 1.3b and fig. 1.7) or to recognize and identify empirical patterns (see the Case of Nathan Gates in chapter 2; and the Learning Cycle example 2 in fig. 6.5). These activities fit within the *Explore* stage of the Learning Cycle.

Alternatively, a Five Practices discussion might occur following a task in which students develop models to *explain* patterns (such as Ms. Nichols's students during the Explaining the Behavior of Water lesson or Mrs. Duncan's students in Learning Cycle example 1, fig. 6.4). In the case of Ms. Nichols's students, the discussion occurred during the *Explain* stage of the Learning cycle. The students produced representations of how the particles might be arranged and might behave in order to account for the patterns they had noted during the prior *Explore* activities. In the case of Mrs. Duncan's class, the Five Practices discussion occurred during the *Apply* stage, following a lesson in which she utilized direct instruction to present the initial explanation for patterns that students had noted in the *Explore* stage. Drawing on this new knowledge, students then had an opportunity to create explanations for a different phenomenon (cement curing) and to present and critique ideas collaboratively.

Given that a Five Practices discussion might be incorporated at various points in the Learning Cycle, teachers must make strategic decisions about when to use such discussions and what additional activity structures to employ throughout a series of lessons. In the next section we present a view of teachers as *instructional designers* and discuss how the Five Practices model and the Learning Cycle might mediate the choices they make as they plan for instruction in their particular classroom contexts.

Teachers as Instructional Designers

The NGSS (Achieve, Inc. 2013), and the National Science Education Standards (NRC 1996) before them, present a vision of science instruction that involves students learning scientific concepts through the sense-making practices of inquiry. However, teachers have much work to do in order to make this vision a reality in U.S. schools. This was shown clearly through the TIMSS Video Study (Roth et al. 2006), in which researchers coded videos from 439 eighth-grade classes from five countries (the U.S., the Netherlands, Czechoslovakia, Australia, and Japan) who participated in the 1999 Trends in International Mathematics and Science Study (TIMSS). The classroom study revealed that, compared to students in other countries, U.S. students spent the greatest amount of time engaged in activities that were not connected to any science concepts (27 percent of tasks) and the least amount of time discussing canonical science ideas (31 percent of public talk time). Overall, the enacted science curriculum in the U.S. engaged students in tasks that lacked coherence between naturally occurring phenomena and the underlying explanations for those phenomena.

How can we begin to address this discrepancy between the ambitious goals reflected in current standards (i.e., the NGSS and National Science Education Standards) and the reality of the enacted curriculum in U.S. schools? A first step is to recognize that teachers are an indispensable part of making the recommended curriculum vision a reality in the particular contexts of their practices (Brown 2009; Brown and Edelson 2003; Fishman and Krajcik 2002; Remillard 2005; Squire et al. 2003). In order for teachers to be able to bring the NGSS alive in their classrooms, they will need to learn how to take on the role of instructional designers. This role involves being aware of the various curricular and professional development resources that are available and considering their appropriateness for specific teaching situations. The ability to assemble resources to the best advantage of a specific set of students at a particular point in time is the essence of what Brown (2009) has referred to as "pedagogical design capacity," and it underlies the instructional design process.

The instructional design process can be informed by models or frameworks that promote instruction consistent with the goals of the NGSS, such as the Five Practices model and the Learning Cycle. While such resources can be useful, they are not an algorithmic solution to the challenges inherent in instructional design. In other words, frameworks can provide guidance, but they don't eliminate the need for teachers to make choices, a point that becomes quite clear when we look carefully at the Learning Cycle and its use in instructional design. One of the strengths of this framework is that teachers may use it to select or design tasks that provide appropriate opportunities for learners (e.g., opportunities to represent and analyze data during the *Explore* stage), while enacting those tasks in the context of activity structures (see fig. 1.6) that may involve different levels of teacher guidance. The Learning Cycle is consistent with a range of instructional approaches (Keys and Bryan 2001; Martin-Hansen 2002), enabling teachers to provide scaffolding appropriate for their students' needs. As such, the Learning Cycle can help teachers to select and sequence a coherent set of lessons without constraining the pedagogical approaches that a teacher uses.

Why does this matter for instructional design? Teachers typically do not have time to address every topic through open-ended inquiry. Inquiry lessons are challenging to enact well, and research has shown that careful guidance throughout inquiry experiences is needed in order for learning to occur (Moreno 2004; Kirschner, Sweller, and Clark 2006). While the sequence of learning experiences within the Learning Cycle is constrained—particularly beginning with the exploration of empirical data and following with the development of explanations or causal models for trends in data—and is consistent with scientific inquiry, use of the Learning Cycle as a pedagogical framework does not mean that teachers must conduct every lesson in an open-ended inquiry format.

Let's revisit example 1 of the Learning Cycle (fig. 6.4) with the goal of understanding the instructional design choices Mrs. Duncan made and what influenced them. Mrs. Duncan's school district provided her with a binder containing information about the expectations for the tenth-grade chemistry course. This binder included a general overview of each unit (including information about how many instructional days should be set aside for each topic), detailed performance objectives, and suggested instructional activities. The first unit within the core curriculum was *Matter and Energy*. It began with a lesson—mostly review from middle school science—about the phases of matter and their general properties. The suggested lesson did not involve asking students to connect molecular structure and behavior to observable or measurable macroscopic properties.

Mrs. Duncan's class was considered a "low level" chemistry class. The majority of her students performed below grade level on tests of mathematics and reading comprehension and did not see themselves as "good" or successful students. Nevertheless, they were enrolled in chemistry because it was a requirement for high school graduation. Mrs. Duncan was concerned about beginning the chemistry course with the suggested lesson about phases of matter because it merely reviewed descriptive knowledge and vocabulary. She worried that this would not be motivating and would convey low expectations, only serving to reinforce her students' general lack of confidence as science learners. She knew that her students had explored phases of matter in their middle school curriculum, and she felt confident that they were familiar with the macroscopic properties of the different phases. She wanted to begin the chemistry class with a Learning Cycle that would enable them to develop some *explanatory* knowledge related to phases of matter and that would include opportunities for them to participate in a Five Practices discussion, as she was committed to using this type of discussion to help her students develop scientific argumentation skills. In particular, she wanted to support their developing abilities to offer and/or demand specific evidence for claims, and she felt that a Five Practices discussion would be an effective activity structure within which to teach and support those skills.

She studied the NGSS and identified the crosscutting concept—*Macroscopic patterns are related to the nature of microscopic and atomic-level structure* (Achieve, Inc. 2013, p. 42)—as a crucial idea that she should return to throughout the chemistry course. She decided to begin with a Learning Cycle that would help her students to build this idea. She felt that once they understood that they could use ideas about molecules and atoms to explain what they could *see,* they'd be more motivated to learn about the structure of atoms as they progressed into more abstract content throughout the course.

Like many science teachers, Mrs. Duncan immediately thought about using water to teach about phases. She reasoned that water is inexpensive, readily available, nontoxic, and can exist in all three phases at normal atmospheric pressure and at temperatures easily obtainable in the laboratory. She did a quick online search and found a laboratory activity designed to help students notice that the volume of water increases when it changes phase from liquid to solid. In this activity, students measured the mass and volume of water samples on day 1, placed the samples in the freezer, and then collected data on the frozen samples on day 2. Mrs. Duncan immediately saw how this pattern—the expansion of water as it freezes—could easily connect to her students' everyday experiences and prior knowledge. Members of their community were always complaining to local authorities about the appalling condition of the roads. In fact, the local newspaper once referred to their city as the "pothole capital of the world." Mrs. Duncan was also confident that at least some of the students would have experienced "exploding" soda cans or water bottles that had been placed in the freezer or left in cars during winter nights. She determined that she could begin the Learning Cycle with an *Engage* scenario that would give students an opportunity to think about how frozen water expands and to share their ideas with the class (see fig. 6.4, Turn & Talk).

Looking at the suggested timeline provided in the core curriculum binder, Mrs. Duncan determined that she could spend four days on this initial Learning Cycle and still be able to cover all of the required topics in unit 1 by the recommended date. She wanted to be sure to provide two full days for Five Practices–related activities: one day for students to work in collaborative groups on a demanding task while she monitored their progress, and one day for the whole-class discussion. She had already determined that this initial activity should highlight the expectation that students develop explanatory (and not merely descriptive) knowledge, so she didn't want to spend too much time on the *Explore* part of the Learning Cycle. She only needed to provide sufficient opportunity for students to familiarize themselves with the data (what they were measuring) and notice the key pattern related to the increase in volume of solid water. Thus, Mrs. Duncan decided to modify the laboratory task so that students could complete the data collection relatively quickly. She decided that she would set up the experiment ahead of time and gather the first day's data. Students would therefore only be responsible for measuring mass and volume of a few samples and contributing their data to the whole-class chart (see fig. 6.4, Small Group Laboratory). In this way, students could quickly have access to enough data to notice the pattern. Mrs. Duncan would also have the opportunity to emphasize her expectation that the class work collaboratively to develop new knowledge.

This left her with three additional days for which to plan. Unlike Ms. Nichols (chapters 3 and 4), she decided not to ask her students to develop their own explanations for the behavior of water. For this first Learning Cycle, Mrs. Duncan wanted to set her students up to be successful at a challenging task. She worried that imagining the molecules of water and inventing ideas related to attractive forces (hydrogen bonds) would not come easily to many of her students, and she wanted to guard against frustrating them. Consequently, she decided to follow up the *Explore* activity with an interactive lecture in which she presented the molecular explanation to the students (fig. 6.4, Interactive Lecture). She planned to provide visual and hands-on representations to support students' engagement and developing understanding. Once the students had learned about the model, Mrs. Duncan felt confident that she could ask them to *apply* their knowledge to a real-world problem with which they were already familiar: the cracking of concrete. Thus, she decided to devote two days of the Learning Cycle to the *Apply* stage, incorporating the challenging task and Five Practices discussion in days 3 and 4 (see fig. 6.4).

In developing the Learning Cycle on the properties of water (fig. 6.4), Mrs. Duncan made many decisions. To begin, she made decisions about *what* students should learn. Specifically, she had to decide, beyond the general topic of phases of matter, what specific patterns and explanatory ideas should comprise the learning goals of the lesson. By drawing on the recommended curriculum (in this case, the NGSS) and the supported curriculum (the core curriculum binder provided by her school district), Mrs. Duncan was able to develop specific learning goals. These included descriptive knowledge (the volume of water increases when it changes from liquid to solid phase) and explanatory knowledge (the H-bonds of solid water hold molecules in a rigid structure that includes more intermolecular space than is found between particles in the liquid). Drawing on the NGSS also enabled Mrs. Duncan to identify an overarching conceptual learning goal: *Macroscopic patterns are related to the nature of microscopic and atomic-level structure.*

Once she had determined what students should learn, Mrs. Duncan had to make decisions about the *order* in which they should consider these patterns and ideas. Here she drew on her understanding of the Learning Cycle framework, beginning in the *Engage* stage by surfacing (and making public) students' existing knowledge related to the phenomenon. Then, building on these early ideas, she decided that the *Explore* stage would provide students with the opportunity to develop descriptive knowledge of the pattern learning goal. Next, the explanatory ideas about H-bonds would be the focus of the *Explain* stage. Finally, she would ask the students to use their new explanatory knowledge to make sense of a new phenomenon in the *Apply* stage.

In addition to decisions about what ideas students should learn and the order in which they should learn them, Mrs. Duncan also made decisions about *how* students should learn. She consulted her curriculum binder and online resources. She selected and designed tasks for the students and identified appropriate activity structures within which to embed those tasks. While selecting and designing tasks, Mrs. Duncan drew upon her understanding of the Five Practices model as well as her knowledge of her own students. She knew that a challenging task involving the production of artifacts was a key component of a Five Practices discussion. She also knew that her students were unlikely to develop the desired explanation for water's properties on their own. Thus, she decided to design a challenging *Apply* task for her students, giving them the opportunity to work collaboratively to use new ideas and build confidence in their capacity to engage in this type of challenging work.

Conclusion

In this chapter we have presented the Learning Cycle framework, which describes a sequence of experiences through which learners can build scientific skills and understandings. We have emphasized that the teacher is the designer of those experiences. Robust models and frameworks like the Five Practices model and the Learning Cycle can mediate the ways in which teachers draw upon curriculum resources to create the enacted curriculum in their classrooms. However, as the case of Mrs. Duncan illustrates, thoughtful science teachers draw flexibly on the Five Practices model and the Learning Cycle, as well as on various curriculum resources, as they make decisions during the process of instructional design in their unique classroom contexts (fig. 6.6). For example, Mrs. Duncan drew upon the NGSS (the "recommended" curriculum), identifying the crosscutting concept *Macroscopic patterns are related to the nature of microscopic and atomic-level structure* (Achieve, Inc. 2013, p. 42) to frame a series of units in the course. She also utilized her district's curriculum binder (the "supported" curriculum) to identify *properties of matter* as the topic of the first lessons in her chemistry course. Having made these decisions, Mrs. Duncan used her understanding of the Learning Cycle to select and design a specific series of learning tasks through which students could develop understanding not only of the properties of water in its different phases but also of the underlying molecular explanation for those properties.

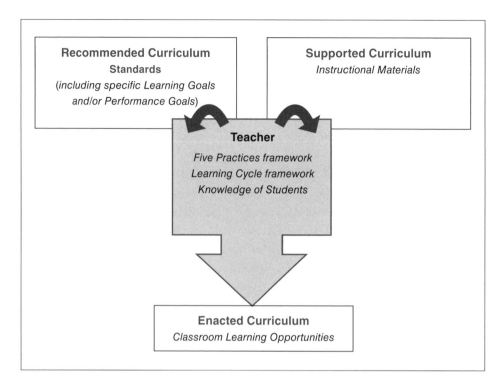

Fig. 6.6. Teachers as designers of the enacted curriculum

In making decisions about what and how students should learn, teachers also draw on deep and complex understandings of their students' experiences and cognitive resources. In Mrs. Duncan's case, consideration of her students' past experiences and concern for their motivation and self-efficacy as science learners led her to develop learning goals that included challenging explanatory knowledge (rather than simply descriptive knowledge about properties of phases). Her understanding of her students also impacted her choice to position the Five Practices discussion during the *Apply* stage of the Learning Cycle. This sequencing enabled Mrs. Duncan to provide careful guidance and support for the students as they grappled with the explanatory knowledge initially (during the interactive lecture, in the *Explain* stage). She concluded the Learning Cycle by providing students with an opportunity to work more independently and to demonstrate mastery of their new explanatory knowledge, thus helping them to develop confidence in their abilities as science learners.

In the next and final chapter of this book, we share examples of how novice teachers have drawn upon the Five Practices model, as well as upon their knowledge of particular students, to design the enacted curriculum. We conclude by summarizing some of the key lessons and promising practices we have learned from their experiences.

CHAPTER 7

Beginning Secondary Science Teachers Use the Model: Lessons Learned

Throughout this book we have provided detailed vignettes that illustrate what classroom instruction and student engagement might look and sound like when teachers utilize the Five Practices model. In this chapter, we step back and reflect on what we have learned from our years of working with preservice and novice teachers who utilized the five practices to design and support science instruction for students in grades 7–12. We begin by providing some background about two contexts in which we have developed insights related to the Five Practices model: our work providing professional development to Knowles Science Teaching Fellows, and our work as instructors in pedagogy courses for preservice science teachers at the University of Pittsburgh. We conclude the chapter with a description of six key lessons learned about challenges and promises of the five practices for supporting ambitious science teaching.

Contexts for Working with Beginning Teachers: Foundation-Supported and University-Based

The Knowles Science Teaching Foundation (KSTF) is a charitable organization established by Janet H. and Harry C. Knowles for the purpose of increasing the number of high-quality science and mathematics teachers in the U.S. Among its other programs, KSTF offers teaching fellowships to a small number of exceptional candidates who are currently obtaining or have recently obtained licensure to teach high school science or mathematics. One of the core principles of KSTF is that "learning to teach requires time, sustained effort, and ongoing support and development throughout a teacher's career" (see http://www .kstf.org/about/mission.html). Consistent with this principle, KSTF offers Teaching Fellows rigorous and continuous support throughout their five-year association with the organization, including two intensive day-and-a-half workshops focused on the Five Practices model. These workshops, which we began providing in 2010, occur during the Science Teaching Fellows' second year with KSTF. They include support for selecting and/or designing cognitively demanding tasks as well as opportunities to learn about the Five

Practices model and to engage in role-play scenarios that involve enactment of aspects of the model. In this chapter, we draw upon our work with a group of Biology Teaching Fellows who shared their struggles and successes related to using the Five Practices model to design and support instruction in their classrooms during 2011–13.

Beginning in 2010, we also integrated opportunities to learn about and utilize the Five Practices model into the core pedagogy courses provided for preservice secondary (grades 7–12) science teachers at the University of Pittsburgh. Teacher candidates receive support to select and design tasks that place high cognitive demand on students and learn about using the Five Practices model to maintain this cognitive demand during instructional enactment. Throughout their two- or three-semester master's-level program, the teacher candidates design, enact, and reflect upon a minimum of three lessons in which they use the five practices to support discussion in their own science classrooms. In this chapter we draw upon our experiences as instructors in the science pedagogy courses as we describe and provide examples of important lessons learned about the Five Practices model.

Lessons Learned

Over the past three years we have provided professional development and training to several cohorts of beginning secondary science teachers in the two contexts described above. As we prepared this book, we took stock of what we have learned through an examination of teacher candidates' written work, direct observations of their practice, and summary evaluations prepared by their field supervisors (where applicable). We created an initial list of "lessons learned" and shared this with the KSTF staff member who designs and supports the professional development program for Teaching Fellows and with the 2010 KSTF Biology Teaching Fellows (with whom we have worked steadily for two years). Their feedback is reflected in the detailed descriptions of lessons learned that we present below.

Lesson One: Plan collaboratively using the Five Practices model

Although many teachers utilize the Five Practices model on their own, the majority of the KSTF fellows with whom we've worked report that collaborative planning with colleagues helps them to design more effective tasks and to anticipate a greater variety of student approaches. The KSTF fellows meet face-to-face only a few times during the year. However, they use group wikis and other web-based sharing tools to support their collaboration related to utilizing the Five Practices model.

CLASSROOM VOICE

"Selecting or designing a task is definitely challenging. When possible, collaborate with others to find or create tasks. Rather than starting from scratch, find a lab or activity you already do or a compelling data set and go from there. It is also helpful to work with others and have a fresh set of eyes when anticipating student responses and possible pitfalls, especially the first time you are trying out a task." (KSTF fellow)

Collaboration with colleagues can support an iterative instructional design process. Often, teachers begin by discussing various ways in which students might approach or make sense of a particular task. Such a discussion serves to highlight features of the task that support the targeted learning goals as well as features that may act as barriers to students' learning, thus providing focus for teachers' revision of the original task. As such, the earliest stages of planning often involve moving back and forth between the design of the task, the identification of learning goals, and the anticipation of student approaches to the task.

The focus of collaborative planning is not limited to consideration of task features and possible student approaches or responses, however. Teachers can also share insights about how to monitor students' work (see fig. 7.1), which ideas to feature in whole-class discussion, and how the selecting and sequencing of different representations and approaches might support attainment of the target learning goals. Such detailed planning can prepare teachers to address issues that arise during instruction, making it more likely that their instructional choices will serve to maintain the cognitive demand of the tasks for learners (Smith and Stein 2011). Moreover, our recent research with more experienced high school science teachers suggests that when teachers plan on their own, their *anticipating* is often limited to consideration of student participation (e.g., how to motivate them to engage in the task, what types of behavior management issues might emerge), or of logistical concerns (e.g., how to distribute materials, what to do if students need more time) (Smith et al. 2013). Thus, at least for some teachers, planning that involves thoughtful anticipation of the ways in which students might make sense of science phenomena may require a productive collaborative context.

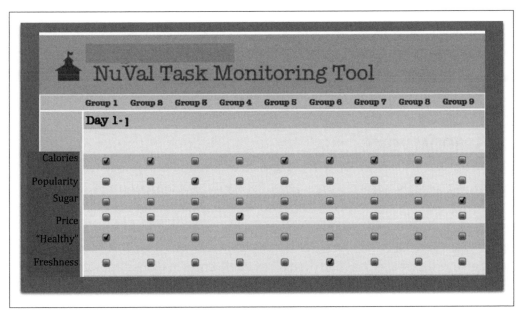

Fig. 7.1. KSTF fellow Brittney Barickman created this monitoring tool on her iPad to help notice and keep track of various components of students' explanations during a group task leading up to a Five Practices discussion. She shared the tool with the other KSTF fellows during the spring 2012 cohort meeting.

Lesson Two: A "student-centered" lesson can still be teacher directed

In guidelines such as the Next Generation Science Standards (NGSS) (Achieve, Inc. 2013), science teachers encounter strong messages about the importance of students having opportunities to recognize and articulate their ideas, participate in learning experiences that serve to challenge and build upon those existing ideas, and actively construct more sophisticated understanding of key science concepts. The novice teachers with whom we work generally find this vision of student-centered instruction appealing; they quickly come to accept and value that learning occurs when students, not teachers, are doing the "thinking work" in the classroom. They believe that students learn through sharing their ideas, listening to and critiquing the ideas of others, and by having others critique their approaches to solving problems.

Through early attempts at enacting this type of lesson—classroom discussion and critique of one another's work—with their own students, many novice teachers quickly learn that it is easy to fall into the trap of allowing students to "hijack" the lesson. In other words, many of our novice teachers have allowed students almost complete autonomy in selecting what ideas to share or follow up on during discussion. At the end of such lessons, the teachers often feel frustration because the target science learning goals were not addressed. They also feel conflicted, believing that stepping in to redirect the discussion or curtail talk that is progressing along unproductive lines is incompatible with their commitment to enacting student-centered instruction.

The Five Practices model has helped our novice teachers to become comfortable with the notion that they can honor and build upon the contributions of their students and still maintain control of the conceptual direction of the class discussion. By carefully anticipating and planning for ways to monitor students' thinking, they are able to notice important features of their students' work. By deliberately selecting and sequencing the opportunities for students to share this thinking, they feel confident that the students' ideas can be leveraged to obtain the target learning goals for the lesson.

CLASSROOM VOICE

"During discussions, teachers often feel a tension between allowing students more control over the conversation and ensuring that the key ideas of the lesson are highlighted and highlighted correctly. The use of the Five Practices eliminates this tension because the teacher plans out the conversation in advance, selecting groups or students to present certain ideas. Because the conversation is well planned, students are able to connect ideas together from different groups and even dissect differences in groups' approaches without you having to intervene in the discussion. This forces students, rather than the teacher, to take ownership of making those connections and asking questions." (KSTF fellow)

Engle (2011) explains that in order for students to be truly engaged in an activity, it is necessary that they have some degree of intellectual authority. Providing students with agency in solving problems encourages students to have ownership over the knowledge produced within the classroom community (Engle and Conant 2002). The caveat with providing authority to students is that there is a very real danger that some learners will act as authorities unto themselves, developing all kinds of unsubstantiated claims without evidentiary warrants (Engle 2011; Engle and Conant 2002). By creating a classroom in which students are responsible for regularly accounting for how their ideas make sense, and by holding students accountable to justifying those ideas with their peers, teachers can create a more student-centered instructional environment that is also accountable to the discipline (Engle 2011, p. 9). Selecting and sequencing the ideas that emerge and posing specific questions to presenters and class members during discussion are strategies that teachers can use to guide the discussion toward targeted learning goals and disciplinary core ideas.

Lesson Three: Don't rush to the destination. Provide sufficient time for students to engage during the journey

Most of the novice teachers with whom we work regularly implement tasks in their classrooms where students work in collaborative groups. During such tasks, the teachers circulate among the groups, usually with two goals in mind. The first goal is to listen carefully and keep track of ideas that emerge. Often, the use of a monitoring tool (see fig. 7.1 for an example) supports this goal. The second goal of the teachers during the collaborative group work is to ensure that students are making progress on the task. Where novices can get tripped up is in how they recognize and support "progress." In this regard, we have identified two "rookie mistakes" that novices tend to make: (1) They seek to move too quickly from students' initial ideas to more sophisticated understandings, often based on hastily made inferences about their thinking; and (2) They attempt to help each individual group develop a complete and correct response or explanation before the class discussion even takes place. We discuss each of these challenges below.

A novice teacher who has used the five practices to anticipate students' work on a task in great detail can easily fall into the trap of jumping to conclusions about students' thinking when she hears a phrase or sees a representation that she anticipated. Imagine, for example, that a teacher planned to use the earlobe genotype task (fig. 4.1) and anticipated that some students would think that because the children all resemble their father, they only inherit genetic information from him (and not from the mother). Now imagine that while monitoring, this teacher heard a student say, "The earlobes all come from the father." Upon hearing this, the novice teacher might infer that the student believes that the mother did not contribute genetic information for this trait to the offspring. Consequently, she might seek to "tutor" the group to ensure that this misconception is eliminated. She might do this by asking a series of questions aimed at leading students to become dissatisfied with what she believes is their model: "What is the process that happens to make a zygote? In this process, how does information go from the mother and father to the offspring? So is it possible for only the father to contribute information? How can you revise your answer to take into account that both parents have to contribute genetic information?"

A teacher more experienced with supporting collaborative group engagement with challenging tasks would be more likely to resist the temptation to make inferences about student thinking or to jump directly into correcting or redirecting students. Instead, a more experienced teacher might use questions to *elicit* additional information about what students think, enabling her to form a more complete picture of their current understanding. For example, a teacher-student exchange during monitoring might sound like this:

Maya:	I think the earlobes all come from the father.
Teacher:	Can you say more about that, Maya?
Maya:	They all look like him?
Teacher:	*(to others in the group)* What do you think about that? If the offspring *look* like the father, does that mean the earlobes *come from* him? Can you unpack that a little bit more?

In this exchange, the teacher did not simply assume that the students believe only the father has passed on genetic material. She asked questions to elicit further clarification of their thinking, seeking to understand whether the students were considering genotype or merely summarizing what they noticed at a phenotypic level. She also built on Maya's response by asking the whole group to consider the connection between a trait "coming from" a parent and the resemblance of those offspring to the parent. In doing so, she *guided* them toward addressing the relationship between genotype and phenotype.

CLASSROOM VOICE

"My colleague took video of my students working on a five practices . . . task, and what struck me most is how much unnecessary hovering and answer-giving I did. That is, I saw students having a great discussion and pushing each other, which was sometimes cut short by me basically coming over and confirming an answer to the point of giving something away. Essentially, rather than letting students struggle and challenge each other, I would do the questioning and confirming of answers for them, thereby reestablishing myself as the center of the classroom. . . . Although the five practices absolutely allow for confidence-building via confirmation of right answers, it's not necessary to confirm right answers every step of the way. Students must be able to construct their own meaning, perhaps with some errors along the way. To this end, some students feel greatly uncomfortable when I don't confirm their ideas every few minutes, and that's something I should probably prepare them for." (KSTF fellow)

The contrast between the novice and experienced teachers in this hypothetical example serves to illustrate the second rookie mistake: Many novices remain with a group during the monitoring process until they are sure that the group is moving toward a complete and correct response or explanation. Consequently, they are likely to engage the group in highly structured questioning that leads them quickly and directly in a particular conceptual direction. Another consequence of novices' reluctance to leave groups to struggle with tasks on their own is that they don't spend equitable amounts of time with all groups. Instead, they get easily caught up in detailed "tutoring sessions" with a few groups and largely ignore others.

In contrast, more experienced teachers deliberately walk away from groups even when they know that their current models or explanations are incomplete. They provide "parting shots" to help students to focus on important features of the data or models in question (see chapter 5) and trust that students will have opportunities to refine their incomplete or incorrect ideas during the whole class discussion that will follow the collaborative group task. Not surprisingly, we have found that persistence and reflection help novice teachers to develop patience with the monitoring process (and avoid turning it into a series of individual tutoring sessions) and trust that the whole-class discussion, when purposefully structured, will lead students to understanding of the targeted learning goals.

CLASSROOM VOICE

"It can be really hard as a teacher to ever let a student or a whole group come to the 'wrong' conclusion or not reach any conclusion at all. While you can provide groups with some scaffolding or extra data during the task that might help them, it is also okay to let questions or confusion come out and be clarified in the discussion. Because you can plan the discussion in advance, you can make sure that the key idea comes up and is explained. It is helpful to make sure all students have access to the data or information from the task so the students who need clarification have everything right in front of them as another group is presenting their understanding based on the data. If you know there is a particularly difficult idea that many groups are struggling with, you might even present them with new data or information to help the whole class understand it better. For example, for one task my students really struggled with the difference between a change in mass and a percent change in mass related to different amounts of transpiration in plants. A few groups understood the concept, but had trouble explaining it. During the discussion, I gave everyone the data set below and asked the simple question, 'Who lost more weight?' As students debated their answers in the discussion, they all came to a better understanding of the difference between a change and a percent change." (KSTF fellow)

	Initial weight (lbs)	Final weight (lbs)	Change in weight (lbs)	% Change in weight
Suzie	120	110	10	8%
Bob	200	188	12	6%

Lesson Four: Include all students in the discussion, even those who are not presenting their work

To include all students in the discussion, teachers must (1) position themselves as facilitators of discussion and not interviewers of individual students; (2) use various "talk moves" in order to help all students access the discussion and hold them accountable for engaging in it; and (3) be explicit with students about expectations for their participation in the whole-class discussion and provide scaffolds to support desired types of engagement.

Teacher as facilitator

We have noticed that novice teachers (or teachers who are novices with respect to facilitating Five Practices discussions) frequently position themselves as "interviewers" of a succession of individual students rather than as facilitators of discussion among students. Consider the following excerpt from a class discussion about the Frog Problem task (see fig. 1.5):

Teacher:	Okay, Marquis, please tell us about your group's work.
Marquis:	We think it was the trematodes that made the deformed frogs.
Teacher:	Why do you think that?
Marquis:	Because only the ponds where there was a high level of deformed frogs also had the trematodes in them. At high levels, I mean.
Teacher:	And can you show us where on your poster we can see the high levels of trematodes?
Marquis:	Here *(points to the table at the bottom of the poster).*
Teacher:	Thank you, Marquis. Does anyone have any comments to make about this group's work?

We would characterize this teacher as an "interviewer" because all of his questions are directed toward Marquis. By posing such pointed questions exclusively to Marquis, this teacher-as-interviewer inadvertently sends the message to the rest of the class that their role is to be audience members, passively sitting and waiting for Marquis to provide answers. The teacher might send a very different message if he had responded to Marquis's claim in the following way:

Marquis:	Because only the ponds where there was a high level of deformed frogs also had the trematodes in them. At high levels, I mean.
Teacher:	So Marquis has made the claim that only the ponds with high trematode levels also had high levels of deformed frogs. *(To the class)* Can you see clearly how this group's poster shows that? Alex, please show the class where you see this on the poster.

This alternative response to Marquis conveys the teacher's expectation that all students should be carefully evaluating whether the group successfully presented evidence to support that claim. By calling on a specific student to interact with Marquis's poster, the teacher also helps the class to understand that any one of them may be invited to contribute to the discussion at any time, thus keeping them "on their toes."

Talk moves

Following a student's initial class presentation, many teachers will simply invite more students to join in using an open-ended prompt such as Marquis' teacher did: "Does anyone have any comments to make about this group's work?" Alternatively, a teacher may be more strategic in how he or she solicits further input from students. For example: "Katie, we just heard Jason tell us that his group decided to use the mean to represent typical plant growth. When I was with your group I heard you talking about mean. Can you tell the class what your group decided and why?" We have found that it is often beneficial to be strategic in soliciting the participation of other class members for two reasons. First, as we saw in the cases of Kendra Nichols and Nathan Gates, it enables the teacher to ensure equitable participation among class members. Second, it helps the teacher to ensure that the discussion progresses along a coherent science "story line."

There are many talk moves that teachers can use to support and shape productive classroom discussions. Here we describe three moves that both novice and pre-service teachers with whom we've worked have found to be particularly useful: (1) teacher revoicing, (2) student revoicing (Chapin, O'Connor, and Anderson 2003, pp. 12–16), and (3) marking.

Teacher Revoicing—Especially in the early stages of getting discussions going in the classroom, student contributions are often difficult to hear and sometimes difficult to understand. Yet all students need to have access to what a student has said if they are expected to think about and comment on it. For this reason, the repeating of part or all of a student response is often a worthwhile move for teachers.

When repeating a student contribution, the teacher must be sure to guard against stripping authorship from the student. If the contribution was too softly stated, the teacher should (after giving the student herself the opportunity to state it more loudly) simply repeat it so that everyone can hear it. If the idea was not well stated or was particularly complex and thus hard for other students in the class to grasp, the teacher might reformulate (but not *change*) the idea to make it more comprehensible. If at all in question, the teacher should also ask the student to respond and verify whether or not the teacher's revoicing was correct. For example:

Andy:	The Moon has phases because it's going around. It's moving. Or, like, it's rotating. No! It's revolving. I always get those confused. It's orbiting the Earth and that's why we see different phases because we can't always see all of the lit-up part from Earth.
Teacher:	Okay, so let me just make sure I understand you. You said that the Moon has different phases because we can't always see all of the lit-up part from our perspective on Earth. And that's because the Moon is revolving, or orbiting, around the Earth. Is that right?
Andy:	Yeah, that's it.

Student Revoicing—Instead of revoicing a student's idea herself, the teacher may ask another member of the class to repeat what a student has just said in his or her own words. This strategy is generally used only when the initial contribution is clear and comprehensible. For example:

Jessica:	We think that all of the plants have an S-shaped growth pattern because all six of ours looked like that. And then we saw that Matt's group and Tevon's group looked the same, too.
Teacher:	Xavier, can you say that again for us? What is Jessica's claim and what is the evidence she's using to support it?

Xavier: She's saying that all of the plants will be S-shaped. Or not the plants, but the growth curve will be S-shaped when you plot it. And . . .

Teacher: And her evidence?

Xavier: Yeah, her evidence is that their group plotted all six plants and they all had an S-shaped graph. Plus the other graphs we saw in the other groups were shaped like an S too.

Again, the idea is to repeat, and to not interpret, evaluate, or critique the response. Students' restating of another student's contribution often serves to mark the idea as being especially important and worth emphasizing. As such, it signals to the author that his or her ideas are being taken seriously, and it puts the rest of the class on notice that they have a second chance to catch up on something really important. Further, use of this strategy is one way that a teacher can communicate and enforce the expectation that all students should be prepared to contribute to the discussion at all times.

Marking—Marking is a strategy that often involves both teacher revoicing and simplification or summary of an especially important point. Teacher talk moves that mark such points can be expected to occur hand in hand with directing attention to salient features of the problem space as discussed in chapter 5.

Students often formulate their ideas at the same moment they are sharing them aloud. As such, many student contributions are complex and involve multiple changes of direction (see Andy's comments about Moon phases and his confusion about rotation and revolution in the example of teacher revoicing above). A teacher can use marking to highlight the important idea or ideas from a particularly lengthy or complex contribution, as we saw in the Case of Kendra Nichols (part 4, lines 364–77):

Fatima: We show how solids are packed together and liquids are not as packed. But the gases are spaced out. We were thinking about how the gas in the plunger—the syringe thing—was able to be compressed. We put it in there and squeezed and the plunger went down. So we think that's because the molecules have all this space or air between them and when you compress them they are moving closer together. But, like a solid you can't push them closer together because they're already as close as possible. That's why solids keep their shape—you can't move the molecules around. We weren't really sure about the liquid molecules. We couldn't compress them—or compress the liquid in the syringe. But we think they're not stuck together like in a solid.

Ms. Nichols: So you are saying that all three of these phases of water are made up of the same kinds of molecules but what is changing is the way they are spaced, how close or far apart they are from one another? Is that right?

Here Ms. Nichols was able to use marking to signal to the class that the important takeaway idea from Fatima's group was that the spacing between water molecules is different in different phases. She further reinforced this idea by recording it in the class summary table (see fig. 4.5).

In another example of marking from the Case of Nathan Gates (lines 185–93), Mr. Gates revoiced Phaedra's contribution and signaled that everyone in the class should attend to it:

Phaedra:	So what happens if there is a really tall plant in the pots compared to all the others?
Mr. Gates:	Yes, what might happen to the mean if there happened to be a really tall plant compared to all the other plants? Would the mean be any different?
Phaedra:	Well, with even just one really tall height, you would have a bigger sum when you add all the heights together, so the mean would be bigger, too.
Mr. Gates:	Let's think about what Phaedra just said. She said if one of the plants were much taller than the rest, the mean would increase. How do you feel about this idea, Mikhail?

Thus, while marking is a talk move that incorporates revoicing, it goes beyond simply making students' ideas *heard* in the classroom. Marking often involves making those ideas more *accessible* by focusing on some elements while eliminating unnecessary detail. In essence, it boosts "signal" and helps to eliminate "noise." Marking also involves directing the entire class's attention to key contributions during discussion.

Explicit expectations and supports for participation

In addition to positioning oneself as a facilitator and strategically using talk moves, teachers can promote students' productive engagement in whole-class discussion by letting them know what is expected and creating tools to support their participation. Such tools might include a diagram or framework for note-taking or a feedback form where each student can list one "idea learned" and one "additional question" he or she has for each presenter. It is crucial for teachers to hold students accountable for their engagement in the class discussion by developing some form of assessment strategy tied to this classroom activity. Many teachers have found that note-taking sheets, peer feedback forms, and exit slips or reflections (such as the one Ms. Nichols assigned at the end of the Behavior of Water discussion; see p. 80) are highly useful means for gathering evidence of students' attentiveness and active engagement during discussion. These written artifacts provide teachers with an opportunity to give feedback and even to assign points based on students' thoughtful contributions to and reflection upon class discussion.

CLASSROOM VOICE

"As with every step in the five practices, I try to do as much work up front to then be able to step back and let students do as much of the work as possible. This includes making sure each group knows what they will be explaining to the rest of the group and what order they will be in, giving all students some sort of task during the discussion (whether that is taking notes, listening for answers to specific questions, or asking the presenters questions), having any student work ready to share, and sometimes even choosing a student facilitator to keep things moving smoothly." (KSTF fellow)

Lesson Five: Be prepared to "stack the deck" with student work samples

Although the work that is shared during a whole-class discussion will ideally be generated by the students in the class, a teacher can introduce a particularly important strategy or model that no one in the class has used by sharing the work of students from other classes (e.g., Boaler and Humphreys 2005; Schoenfeld 1998) or by offering one of his or her own for the class to consider (e.g., Baxter and Williams 2010). This will ensure that the key ideas needed to move the discussion in a fruitful direction are made public.

For example, in preparing for her lesson on the behavior of water, Ms. Nichols created a complete and correct representation of solid, liquid, and gaseous water that incorporated the features she had identified (see fig. 3.3). While she decided not to share this model with the class, such a model could have been shared following group B's discussion of heat. The model could be attributed to a group of students in another class or the previous year, one downloaded from the web, or even one created by the teacher. The key to effectively using a contribution that was not produced by someone in the class is engaging the class in a discussion regarding how the new ideas are similar to or different from the others that have been discussed, whether the differences are important, and what the ideas add to the class's understanding. In the case of the Behavior of Water task, Ms. Nichols' students could have been asked to examine the new model and consider whether it accurately depicted all of the key features that had been discussed and what, if anything, they would change.

As we indicated in chapter 3, the practice of anticipation involves carefully considering (1) the key features that must be present for a complete and correct experimental design, explanation, or representation; (2) the challenges that students are likely to encounter and/or the misconceptions that might surface as they engage in the task; and (3) how to respond to the work students produce that may or may not address the identified features. It is during this phase of work that the teacher should carefully consider what models, solutions, or representations are critical in order to have a productive discussion that highlights the key disciplinary ideas to be learned. The teacher can then find or create these key models or representations so they will be able to insert them into the discussion as needed if they do not naturally occur. Toward this end, we encourage teachers to archive the work produced by students each year so that it can be utilized in subsequent years. For example, if Mr. Gates took photographs of the graphs his students produced (see chapter 2), he would then be able to share particular graphs that make key features salient the following year if such graphs do not emerge.

In addition to considering complete and correct responses, teachers should also consider being ready to share responses that may raise important issues that they want to make sure are "on the table" for consideration. For example, Mr. Gates may want to share the drawing of plants produced by group 1 (see fig. 2.4). This representation does not help answer the question regarding typicality, but it does help to show that some representations make it challenging to identify patterns and correlations in the data. In the genotype task, discussed in chapter 4 (fig. 4.1), the teacher thought it unlikely that her students would propose a solution in which the father carried the heterozygous (Ff) genotype. Knowing that this is an important possibility for them to consider, the teacher created a fictitious student response in which the father carried this Ff genotype (fig 4.4). She kept this work ready to present during the whole-class discussion as "work from another class." In doing so, the teacher positioned herself to provide students with an opportunity to engage in further discussion about the underlying concept of monogenic inheritance.

> ## CLASSROOM VOICE
> "I find it very helpful when time allows to have the discussion a day *after* I have a good idea of what each group has done. This gives me time to think about how best to sequence their work and to find work samples from another class or make them myself if something is missing." (KSTF fellow)

Lesson Six: The Five Practices model can help teachers support students' developing confidence as science learners in equitable ways

Many of the KSTF Teaching Fellows work with diverse students, including many English Language Learners. The fellows have shared that these students, many of whom have not been successful in traditional science classes, are quite motivated to engage in Five Practices discussions. Such discussions involve tasks for which multiple approaches are valued, and students with diverse expertise are energized by the opportunity to make real contributions to the group's thinking. Conversely, many students who excel at highly structured tasks find collaborative group work and Five Practices discussions to be frustrating because a clear pathway to "success" or the "right answer" is not provided. Some Teaching Fellows have noted that turning the tables in this way can actually serve to promote equity in their classrooms, giving greater voice to students who often lack confidence in the worth of their contributions and providing supportive challenges through which traditionally successful students can grow.

> ## CLASSROOM VOICE
> "While some groups really took off, some definitely struggled as well, especially in my larger classes. Some students who are very successful in a traditional note-taking class had a hard time not being told exactly what to do. Nonetheless, they all made progress and I think they would improve if we did this type of work more often." (KSTF fellow)

Lotan (2006) argues that to build equitable classrooms, teachers must engage in curriculum design that takes into consideration students' academic skills and previous experiences, their linguistic variability, and their intellectual diversity. Teachers must design tasks that have multiple entry points and consider carefully the different ways a task can be approached and the tools that might support students' effort to make sense of the task. The KSTF fellows have found that *anticipating* during the planning process with their specific students in mind enables them to meet their varied needs through carefully structured collaborative group work. For example, teachers who want to use a data representation and analysis task as the basis of a Five Practices discussion will carefully consider the mathematical and reading abilities of their students when developing or selecting data sets. Such teachers ask each group to consider specific data sets based on the affordances and challenges they have deliberately engineered into them as well as on their knowledge of individual students' strengths and weaknesses. The approach of providing a variety of data sets or representations during a task serves to meet students' individual needs while also creating an authentic context within which they must communicate. Because individual groups examine slightly different information, each is responsible for helping the others to see important features in the data to which they had access. This in turn provides an opportunity for students to publicly demonstrate their competence, serving to build their confidence as science learners (Lotan 2003).

CLASSROOM VOICE

"Along with the open-endedness of the task, this is one of the first projects I have done where it is really okay for groups to be doing many different things and working at different paces. Their different approaches are leading them to explore different concepts and, by the end, some groups will have seen data that other groups did not. They will present on different things. And it's okay! While that seemed a little scary to me at first, it is actually really liberating. As long as you think about how students will still achieve the main learning goals and you are prepared with support for the different paths groups might go down (either with different sets of data or questions to guide them a certain way), then there is a lot to be gained from letting them work out what they think is important as a group." (KSTF fellow)

Conclusion

Our collaboration with many thoughtful and dedicated teachers over the last three years has helped to shape our understanding of the challenges and promises associated with the Five Practices model. It is our hope that the resources provided in this book will help teachers at all levels of experience to design and enact instruction that makes the ambitious vision for science instruction portrayed in the NGSS a reality in our nation's schools.

REFERENCES

Achieve, Inc. *The Next Generation Science Standards*. http://www.nextgenscience.org. Washington, D.C.: Achieve, Inc. 2013.

Atkin, J. Myron, and Robert Karplus. "Discovery or Invention?" *The Science Teacher* 29, no. 5 (1962): 45–51.

Baxter, Juliet A., and Steven Williams. "Social and Analytic Scaffolding in Middle School Mathematics: Managing the Dilemma of Telling." *Journal of Mathematics Teacher Education* 13, no. 1 (2010): 7–26.

Boaler, Jo, and Cathy Humphreys. *Connecting Mathematical Ideas: Middle School Video Cases to Support Teaching and Learning*. Portsmouth, N.H.: Heinemann, 2005.

Boaler, Jo, and Megan Staples. "Creating Mathematical Futures Through an Equitable Teaching Approach: The Case of Railside School." *The Teachers College Record* 110, no. 3 (2008): 608–45.

Bransford, John D., Ann L. Brown, and Rodney R. Cocking (eds.). *How People Learn: Brain, Mind, Experience, and School*. Washington, D.C.: National Academy Press, 2000.

Brown, Matthew W. "The Teacher-Tool Relationship: Theorizing the Design and Use of Curriculum Materials." In *Mathematics Teachers at Work: Connecting Curriculum Materials and Classroom Instruction*, edited by Janine T. Remillard, Beth A. Herbel-Eisenmann, and Gwendolyn M. Lloyd, pp. 17–36. New York: Routledge, Taylor and Francis, 2009.

Brown, Matthew, and Daniel Edelson. "Teaching as Design: Can We Better Understand the Ways in Which Teachers Use Materials So We Can Better Design Materials to Support Their Changes in Practice?" *Design Brief* from The Center for Learning Technologies in Urban Schools, 2003. http://www.inquirium.net/people/matt/teaching_as_design-Final.pdf (accessed 20 May 2013).

Bybee, Roger W. *BSCS 5E Instructional Model*. Colorado Springs, Colo.: BSCS, 2002.

Bybee, Roger W., Joseph A. Taylor, April Gardner, Pamela Van Scotter, Janet Carlson Powell, Anne Westbrook, and Nancy Landes. *The BSCS 5E Instructional Model: Origins, Effectiveness, and Applications*. Colorado Springs, Colo.: BSCS, 2006. http://www.bscs.org/sites/default/files/_legacy/BSCS_5E_Instructional_Model-Full_Report.pdf (accessed 20 May 2013).

Cartier, Jennifer L., Cynthia M. Passmore, Jim Stewart, and John P. Willauer. "Involving Students in Realistic Scientific Practice: Strategies for Laying Epistemological Groundwork." In *Everyday Matters in Science and Mathematics: Studies of Complex Classroom Events*, edited by Ricardo Nemirovsky, Ann S. Rosebery, Jesse Solomon, and Beth Warren, pp. 267–98. Mahwah, N.J.: Lawrence Erlbaum Associates, 2005.

Cazden, Courtney B. *Classroom Discourse: The Language of Teaching and Learning*. Portsmouth, N.H.: Heinemann, 2001.

Chapin, Suzanne H., Catherine O'Connor, and Nancy Canavan Anderson. *Classroom Discussions: Using Math Talk to Help Students Learn: Grades 1–6*. Sausalito, Calif.: Math Solutions, 2003.

Davis, Elizabeth A., Debra Petish, and Julie Smithey. "Challenges New Science Teachers Face." *Review of Educational Research* 76, no. 4 (2006): 607–51.

DeBoer, George E. *A History of Ideas in Science Education: Implications for Practice*. New York: Teachers College Press, 1991.

Dewey, John. *The Child and the Curriculum*. Chicago: University of Chicago Press, 1902.

Doyle, Walter. "Academic Work." *Review of Educational Research* 53, no. 2 (1983): 159–99.

Drake, Corey, and Miriam Gamoran Sherin. "Practicing Change: Curriculum Adaptation and Teacher Narrative in the Context of Mathematics Education Reform." *Curriculum Inquiry* 36, no. 2 (2006): 153–87.

Eisenkraft, Arthur. "Expanding the 5E Model: A Proposed 7E Model Emphasizes 'Transfer of Learning' and the Importance of Eliciting Prior Understanding." *The Science Teacher* 70, no. 6 (2003): 56–59.

Engle, Randi A. "The Productive Disciplinary Engagement Framework: Origins, Key Concepts and Developments." In *Design Research on Learning and Thinking in Educational Settings: Enhancing Intellectual Growth and Functioning*, edited by David Yun Dai, pp. 161–200. London: Taylor & Francis, 2011.

Engle, Randi A., and Faith R. Conant. "Guiding Principles for Fostering Productive Disciplinary Engagement: Explaining an Emergent Argument in a Community of Learners Classroom." *Cognition and Instruction* 20, no. 4 (2002): 399–483.

Fishman, Barry J., and Joseph Krajcik. "What Does It Mean to Create Sustainable Science Curriculum Innovations? A Commentary." *Science Education* 87, no. 4 (2003): 564–73.

Hatano, Giyoo, and Kayoko Inagaki. "Sharing Cognition through Collaborative Comprehension Activity." In *Perspectives on Socially Shared Cognition*, edited by Lauren B. Resnick, John M. Levine, and Stephanie D. Teasley, pp. 331–480. Washington, D.C.: American Psychological Association, 1991.

Hiebert, James, and James W. Stigler. "Improving Mathematics Teaching." *Educational Leadership* 61, no. 5 (2004): 12–17.

Hiebert, James, Anne K. Morris, Dawn Berk, and Amanda Jansen. "Preparing Teachers to Learn from Teaching." *Journal of Teacher Education* 58, no. 1 (2007): 47–61.

Institute for Research on Teaching. *Matter and Molecules* curriculum. East Lansing, Mich.: College of Education, Michigan State University, 1998.

Keys, Carolyn W., and Lynn A. Bryan. "Co-Constructing Inquiry-Based Science with Teachers: Essential Research for Lasting Reform." *Journal of Research in Science Teaching* 38, no. 6 (2001): 631–45.

Kirschner, Paul A., John Sweller, and Richard E. Clark. "Why Minimal Guidance during Instruction Does Not Work: An Analysis of the Failure of Constructivist, Discovery, Problem-Based, Experiential, and Inquiry-Based Teaching." *Educational Psychologist* 41, no. 2 (2006): 75–86.

Lampert, Magdalene. *Teaching Problems and the Problems of Teaching.* New Haven, Conn.: Yale University Press, 2001.

Lawson, Anton E. "Using the Learning Cycle to Teach Biology Concepts and Reasoning Patterns." *Journal of Biological Education* 35, no. 4 (2001): 165–69.

Lawson, Anton E., Michael R. Abraham, and John W. Renner. "A Theory of Instruction: Using the Learning Cycle to Teach Science Concepts and Thinking Skills." Monograph #1, National Association for Research in Science Teaching, 1989.

Lotan, Rachel A. "Group-Worthy Tasks." *Educational Leadership* 60, no. 6 (2003): 72–75.

_____. "Teaching Teachers to Build Equitable Classrooms." *Theory Into Practice* 45, no. 1 (2006): 32–39.

Marshall, Jeff C. "4E x 2 Instructional Model: Promoting Stronger Teaching and Deeper Conceptual Understanding." Paper presented at the School Science and Mathematics Association Annual Convention, Indianapolis, Ind., November 15–17, 2007.

Marshall, Jeff C., Bob Horton, and Julie Smart. "4E x 2 Instructional Model: Uniting Three Learning Constructs to Improve Praxis in Science and Mathematics Classrooms." *Journal of Science Teacher Education* 20, no. 6 (2009): 501–16.

Martin-Hansen, Lisa. "Defining Inquiry: Exploring the Many Types of Inquiry in the Science Classroom." *Science Teacher* 69, no. 2 (2002): 34–37.

Mehan, Hugh. *Learning Lessons: Social Organization in the Classroom.* Cambridge, Mass.: Harvard University Press, 1979.

Moreno, Roxana. "Decreasing Cognitive Load for Novice Students: Effects of Explanatory Versus Corrective Feedback in Discovery-Based Multimedia." *Instructional Science* 32, no. 1–2 (2004): 99–113.

National Research Council (NRC). *A Framework for K–12 Science Education: Practices, Crosscutting Concepts, and Core Ideas*, edited by Helen Quinn, Heide Schweingruber, and Thomas Keller. Washington, D.C.: National Academies Press, 2012.

_____. *National Science Education Standards*. Washington, D.C.: National Academy Press, 1996.

Newton, Douglas P. *Talking Sense in Science: Helping Children Understand Through Talk*. London: Routledge Falmer, 2002.

O'Sullivan, Christine Y., and Andrew R. Weiss. *Student Work and Teacher Practices in Science: A Report on What Students Know and Can Do*. Washington, D.C.: U.S. Department of Education Office of Educational Research and Improvement, 1999.

Remillard, Janine T. "Examining Key Concepts in Research on Teachers' Use of Mathematics Curricula." *Review of Educational Research* 75, no. 2 (2005): 211–46.

Roth, Kathleen J., Stephen L. Druker, Helen E. Garnier, Meike Lemmens, Catherine Chen, Takako Kawanaka, Dave Rasmussen, et al. "Teaching Science in Five Countries: Results From the TIMSS 1999 Video Study." Statistical Analysis Report, NCES 2006–2011. National Center for Education Statistics. Washington D.C.: U.S. Government Printing Office, 2006.

Schoenfeld, Alan H. "Toward a Theory of Teaching-in-Context." *Issues in Education* 4, no. 1 (1998): 1–94.

Schwab, Joseph J. "The Teaching of Science as Inquiry." *Bulletin of the Atomic Scientists* 14 (1958): 374–79.

Smith, Margaret S., and Mary Kay Stein. *5 Practices for Orchestrating Productive Mathematics Discussions*. Reston, Va., and Thousand Oaks, Calif.: National Council of Teachers of Mathematics and Corwin Press, 2011.

Smith, Margaret S., Jennifer L. Cartier, Samuel L. Eskelson, and Danielle K. Ross. "Planning and Teaching: An Investigation of Two Teachers' Participation in Collaborative Lesson Planning Activities and the Impact of These Activities on Their Instruction." Paper presented at the annual meeting of the American Educational Research Association, San Francisco, Calif., April 27–May 1, 2013.

Smith, Margaret S., Victoria Bill, and Elizabeth K. Hughes. "Thinking Through a Lesson Protocol: A Key for Successfully Implementing High-Level Tasks." *Mathematics Teaching in the Middle School* 14, no. 3 (2008): 132–38.

Squire, Kurt D., James G. MaKinster, Michael Barnett, April Lynn Luehmann, and Sasha L. Barab. "Designed Curriculum and Local Culture: Acknowledging the Primacy of Classroom Culture." *Science Education* 87, no. 4 (2003): 468–89.

Stein, Mary Kay, and Suzanne Lane. "Instructional Tasks and the Development of Student Capacity to Think and Reason: An Analysis of the Relationship between Teaching and Learning in a Reform Mathematics Project." *Educational Research and Evaluation* 2, no. 1 (1996): 50–80.

Stein, Mary Kay, Barbara W. Grover, and Marjorie Henningsen. "Building Student Capacity for Mathematical Thinking and Reasoning: An Analysis of Mathematical Tasks Used in Reform Classrooms." *American Educational Research Journal* 33, no. 2 (1996): 455–88.

Stewart, James, Jennifer L. Cartier, and Cynthia M. Passmore. "Developing Understanding Through Model-Based Inquiry." In *How Students Learn: History in the Classroom*, edited by M. Suzanne Donovan and John D. Bransford, pp. 515–65. Washington, D.C.: National Academies Press, 2005.